Neuro-Fuzzy Control
of Industrial Systems with
Actuator Nonlinearities

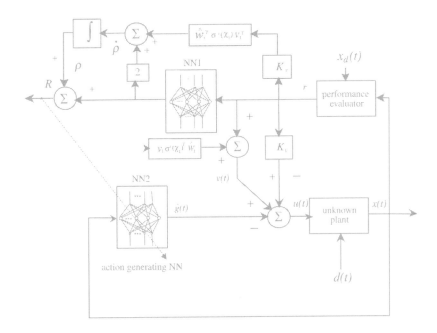

FRONTIERS IN APPLIED MATHEMATICS

The SIAM series on Frontiers in Applied Mathematics publishes monographs dealing with creative work in a substantive field involving applied mathematics or scientific computation. All works focus on emerging or rapidly developing research areas that report on new techniques to solve mainstream problems in science or engineering.

The goal of the series is to promote, through short, inexpensive, expertly written monographs, cutting edge research poised to have a substantial impact on the solutions of problems that advance science and technology. The volumes encompass a broad spectrum of topics important to the applied mathematical areas of education, government, and industry.

EDITORIAL BOARD

BOOKS PUBLISHED IN FRONTIERS IN APPLIED MATHEMATICS

Lewis, F. L.; Campos, J.; and Selmic, R., *Neuro-Fuzzy Control of Industrial Systems with Actuator Nonlinearities*

Bao, Gang; Cowsar, Lawrence; and Masters, Wen, editors, *Mathematical Modeling in Optical Science*

Banks, H. T.; Buksas, M. W.; and Lin, T., *Electromagnetic Material Interrogation Using Conductive Interfaces and Acoustic Wavefronts*

Oostveen, Job, *Strongly Stabilizable Distributed Parameter Systems*

Griewank, Andreas, *Evaluating Derivatives: Principles and Techniques of Algorithmic Differentiation*

Kelley, C. T., *Iterative Methods for Optimization*

Greenbaum, Anne, *Iterative Methods for Solving Linear Systems*

Kelley, C. T., *Iterative Methods for Linear and Nonlinear Equations*

Bank, Randolph E., *PLTMG: A Software Package for Solving Elliptic Partial Differential Equations. Users' Guide 7.0*

Moré, Jorge J. and Wright, Stephen J., *Optimization Software Guide*

Rüde, Ulrich, *Mathematical and Computational Techniques for Multilevel Adaptive Methods*

Cook, L. Pamela, *Transonic Aerodynamics: Problems in Asymptotic Theory*

Banks, H. T. , *Control and Estimation in Distributed Parameter Systems*

Van Loan, Charles, *Computational Frameworks for the Fast Fourier Transform*

Van Huffel, Sabine and Vandewalle, Joos, *The Total Least Squares Problem: Computational Aspects and Analysis*

Castillo, José E., *Mathematical Aspects of Numerical Grid Generation*

Bank, R. E., *PLTMG: A Software Package for Solving Elliptic Partial Differential Equations. Users' Guide 6.0*

McCormick, Stephen F., *Multilevel Adaptive Methods for Partial Differential Equations*

Grossman, Robert, *Symbolic Computation: Applications to Scientific Computing*

Coleman, Thomas F. and Van Loan, Charles, *Handbook for Matrix Computations*

McCormick, Stephen F., *Multigrid Methods*

Buckmaster, John D., *The Mathematics of Combustion*

Ewing, Richard E., *The Mathematics of Reservoir Simulation*

Neuro-Fuzzy Control
of Industrial Systems with
Actuator Nonlinearities

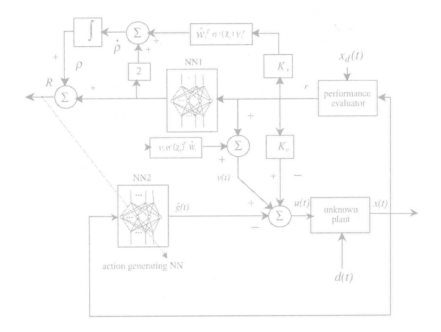

F. L. Lewis
University of Texas at Arlington
Arlington, Texas

J. Campos
Montavista Software, Inc.
Irving, Texas

R. Selmic
Signalogic, Inc.
Dallas, Texas

siam.

Society for Industrial and Applied Mathematics
Philadelphia

Copyright © 2002 by the Society for Industrial and Applied Mathematics.

10 9 8 7 6 5 4 3 2 1

Library of Congress Cataloging-in-Publication Data
Lewis, Frank L.
    Neuro-fuzzy control of industrial systems with actuator nonlinearities / F.L. Lewis, J. Campos, R. Selmic.
        p. cm. – (Frontiers in applied mathematics)
    Includes bibliographical references and index.
    ISBN 0-89871-505-9
        1. Adaptive control systems. 2. Fuzzy systems. 3. Nonlinear theories.
        I. Campos, J. II. Selmic, R. III. Title. IV. Series.

TJ217 .L47 2002
629.8'36–dc21   2002017737

3COM and Etherlink are registered trademarks of 3COM Corporation.

CompactPCI is a registered trademark of PCI Industrial Computers Manufacturers Group (PICMG).

IBM is a registered trademark of IBM Corporation in the United States.

IndustryPack is a trademark owned by SBS Technologies, Inc., registered with the U.S. Patent and Trademark office.

LabVIEW and National Instruments are trademarks of National Instruments Corporation.

Motorola is a registered trademark of Motorola, Inc.

Novell and NE2000 are trademarks or registered trademarks of Novell, Inc., in the United States and other countries.

NuBus is a trademark of Texas Instruments.

Pentium is a registered trademark of Intel Corporation.

SNX, VRTX, and VRTXsa are trademarks or registered trademarks of Mentor Graphics.

Systran is a registered trademark of Systran Corporation.

Windows and Visual Basic are trademarks or registered trademarks of Microsoft Corporation in the United States and other countries.

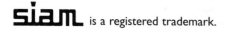 is a registered trademark.

This book is for Galina, for her beauty, strength, and peace.

To Helena, the love of my life.

To Sandra.

Contents

Preface

Since 1960, modern feedback control systems have been responsible for major successes in aerospace, vehicular, industrial, and defense systems. Modern control techniques were developed using frequency domain, state-space, and nonlinear systems techniques that were responsible for very effective flight control systems, space system controllers, ship and submarine controllers, vehicle engine control systems, and industrial manufacturing and process controllers. Recently, the increasing complexity of manmade systems has placed severe strains on these modern controller design techniques. More stringent performance requirements in both speed of response and accuracy have challenged the limits of modern control. Different operation regimes require controllers to be adaptive and have switching and learning capabilities. Tolerance to faults and failures requires controllers to have aspects of intelligent systems. Complex systems have unknown disturbances, unmodeled dynamics, and unstructured uncertainties. The actuators that drive modern systems can be hydraulic, electrical, pneumatic, and so on, and have severe nonlinearities in terms of friction, deadzones, backlash, or time delays.

So-called "intelligent control systems" are modeled after biological systems and human cognitive capabilities. They possess learning, adaptation, and classification capabilities that hold out the hope of improved control for today's complex systems. In this book we explore improved controller design through two sorts of intelligent controllers, those based on neural networks and those based on fuzzy logic systems. Neural networks capture the parallel processing and learning capabilities of biological nervous systems, and fuzzy logic captures the decision-making capabilities of human linguistic and cognitive systems.

This book brings neural networks (NNs) and fuzzy logic (FL) together with dynamical control systems. The first chapter provides background on neural networks and fuzzy logic systems, while the second provides background on dynamical systems, stability theory, and industrial actuator nonlinearities including friction, deadzone, and backlash. Several powerful modern control approaches are used in this book for the design of intelligent controllers. Thorough development, rigorous stability proofs, and simulation examples are presented in each case. In Chapter 3, we use feedback linearization to design an NN controller for systems with friction. Chapter 4 uses function inversion to provide NN and FL controllers for systems with deadzones. Chapter 5 uses dynamic inversion to design NN controllers for actuators with backlash. Chapter 6 uses backstepping to design an FL controller for a vehicle active suspension system. In Chapter 7, we discuss some aspects of hierarchical supervisory control by developing an adaptive critic controller. Systems with communication time delays are treated in Chapter 8.

An important aspect of any control system is its implementation on actual industrial systems. Therefore, in Chapter 9 we develop the framework needed to implement intelligent control systems on actual systems. The appendices contain the computer code needed to build intelligent controllers for real-time applications.

This book has been written for students in a college curriculum, for practicing engineers in industry, and for university researchers. Detailed derivations and computer simulations show how to understand the controllers as well as how to build them. The step by step derivations allow the controllers to be modified for particular specialized situations.

In several cases the developments include the work of Ph.D. students who are not listed as book authors. In these cases, the chapters contain authorship information assigning due credit to these former students. The bulk of this research was supported by the Army Research Office under grant DAAD19-99-1-0137 and by the State of Texas under grant 003656-027. We would like to thank our ARO Program Directors, Linda Bushnell and, more recently, Hua Wang, for their leadership and direction.

F. L. Lewis

J. Campos

R. Selmic

Chapter 1

Background on Neural Networks and Fuzzy Logic Systems

In this chapter, we provide a brief background on neural networks (NNs) and fuzzy logic (FL) systems, covering the topics that will be needed for control system design using these techniques. Included are NN and FL system structures, learning, and the approximation property. We present different types of NN systems and show how NNs and FL systems are related. In fact, FL systems are NNs with a special structure.

Both NNs and FL systems belong to a larger class of systems called *nonlinear network structures* (Lewis (1999)) that have some properties of extreme importance for feedback control systems. These are made up of multiple interconnected nodes and can learn by modifying the weights interconnecting the nodes. This process is called *tuning* the weights. Changing the weights changes the knowledge stored in the network; in fact, the values of the weights represent the *memory* of the network. NNs are a special sort of nonlinear network that are modeled on the nervous systems of biological systems. FL systems are a special sort that are modeled after the linguistic and reasoning abilities of humans. It is fascinating that there are such close connections between these two sorts of nonlinear networks, as we reveal in this chapter.

1.1 Neural networks

NNs (Haykin (1994), Kosko (1992)) are closely modeled on biological processes for information processing, specifically the nervous system and its basic unit, the *neuron*. The neuron receives multiple signals from other neurons through its dendrites, each signal multiplied by a weighting coefficient. These signals are added in the cell body or soma, and when the composite signal reaches a threshold value, a signal known as the action potential is sent through the axon, which is the neuron's output channel. More detail on the material in this chapter may be found in Lewis, Jagannathan, and Yesildirek (1999).

1.1.1 Two-layer neural network topology and tuning

A mathematical model of an NN is shown in Figure 1.1.1. This NN has two layers of adjustable weights and is known as a *two-layer NN*. The values x_k are the NN inputs and y_i its outputs. Function $\sigma(.)$ is a nonlinear *activation function* contained in the *hidden layer* (the middle layer of nodes) of the NN. The hidden-layer weights are v_{jk} and the output-layer

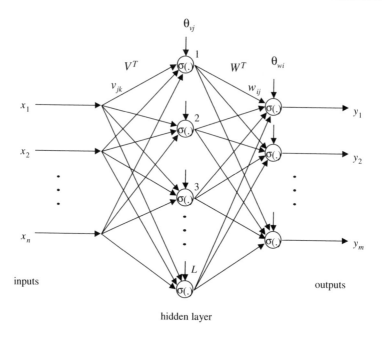

Figure 1.1.1. *Two-layer NN.*

weights are w_{ij}. The hidden-layer thresholds are θ_{vj} and the output-layer thresholds are θ_{wi}. The number of hidden-layer neurons is L.

Neural network structure

A mathematical formula describing the NN is given by

$$y_i = \sum_{j=1}^{L}\left[w_{ij}\sigma\left(\sum_{k=1}^{n} v_{jk}x_k + \theta_{vj}\right) + \theta_{wi}\right], \quad i = 1, 2, \ldots, m. \qquad (1.1.1)$$

We can streamline this by defining the weight matrices

$$V^T = \begin{bmatrix} \theta_{v1} & v_{11} & \cdots & v_{1n} \\ \theta_{v2} & v_{21} & \cdots & v_{2n} \\ \vdots & \vdots & & \vdots \\ \theta_{vL} & v_{L1} & \cdots & v_{Ln} \end{bmatrix}, \qquad W^T = \begin{bmatrix} \theta_{w1} & w_{11} & \cdots & w_{1L} \\ \theta_{w2} & w_{21} & \cdots & w_{2L} \\ \vdots & \vdots & & \vdots \\ \theta_{wm} & w_{m1} & \cdots & w_{mL} \end{bmatrix}, \qquad (1.1.2)$$

which contain the thresholds as the first columns. Then the NN can be written as

$$y = W^T\sigma(V^Tx), \qquad (1.1.3)$$

where the output vector is $y = [y_1 \quad y_2 \quad \cdots \quad y_m]^T$.

Note that, due to the fact that the thresholds appear as the first columns of the weight matrices, one must define the input vector augmented by a "1" as

$$x = \begin{bmatrix} 1 & x_1 & x_2 & \cdots & x_n \end{bmatrix}^T,$$

for then one has the jth row of $V^T x$, given by

$$\begin{bmatrix} \theta_{vj} & v_{j1} & v_{j2} & \cdots & v_{jn} \end{bmatrix} \begin{bmatrix} 1 \\ x_1 \\ x_2 \\ \vdots \\ x_n \end{bmatrix} = \theta_{vj} + \sum_{k=1}^{n} v_{jk} x_k,$$

as required. Likewise, the $\sigma(.)$ used in (1.1.3) is the augmented hidden-layer function vector, defined for a vector $w = [w_1 \quad w_2 \quad \cdots \quad w_L]^T$ as

$$\sigma(w) = \begin{bmatrix} 1 & \sigma(w_1) & \cdots & \sigma(w_L) \end{bmatrix}^T.$$

The computing power of the NN comes from the facts that the activation functions $\sigma(.)$ are nonlinear and that the weights W and V can be modified or tuned through some learning procedure. Typical choices for the activation function appear in Figure 1.1.2. It is important to note that the thresholds θ_{vj} shift the activation function at node j while the hidden-layer weights v_{jk} scale the function.

Weight tuning

The NN learns by modifying the weights through tuning. The NN weights may be tuned using several techniques. A common weight-tuning algorithm is the *gradient algorithm*

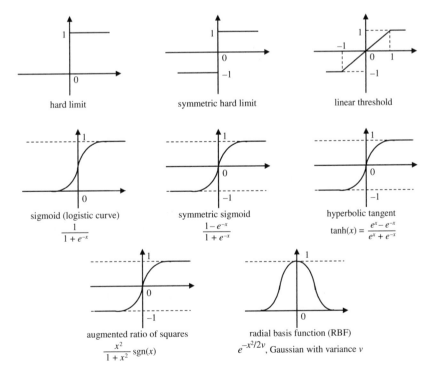

Figure 1.1.2. *Some common choices for the activation function.*

based on the *backpropagated error* (Werbos (1989))

$$W_{t+1} = W_t + F\sigma(V_t^T x_d)E_t^T,$$
$$V_{t+1} = V_t + Gx_d(\sigma_t'^T W_t E_t)^T. \tag{1.1.4}$$

This is a discrete-time tuning algorithm where the time index is t. The desired NN output in response to the reference input $x_d \in R^n$ is prescribed as $y_d \in R^m$. The NN output error at time t is $E_t = y_d - y_t$, where $y_t \in R^m$ is the actual NN output at time t. F, G are weighting matrices selected by the designer that determine the speed of convergence of the algorithm.

The term $\sigma'(.)$ is the derivative of the activation function $\sigma(.)$, which may be easily computed. For the sigmoid activation function, for instance, the hidden-layer output gradient is

$$\sigma' \equiv \text{diag}\{\sigma(V^T x_d)\}[I - \text{diag}\{\sigma(V^T x_d)\}],$$

where diag{.} is a diagonal matrix with the indicated elements on the diagonal.

The continuous-time version of backpropagation tuning is given as

$$\bar{W} = F\sigma(V^T x_d)E^T,$$
$$\bar{V} = Gx_d(\sigma'^T W E)^T. \tag{1.1.5}$$

A simplified weight-tuning technique is the *Hebbian* algorithm, which is given in continuous time as

$$\bar{W} = F\sigma(V^T x)E^T,$$
$$\bar{V} = Gx(\sigma(V^T x))^T. \tag{1.1.6}$$

In Hebbian tuning, it is not necessary to compute the derivative σ'. Instead, the weights in each layer are tuned based on the outer product of the input and output signals of that layer. It is difficult to guarantee NN learning performance using Hebbian tuning.

1.1.2   Single-layer neural networks

We shall see that multilayer NNs have some advantages in feedback control design. However, it can be advantageous to use one-layer NNs in some applications. There are many different sorts of one-layer NNs, and, indeed, we shall see that FL systems can be considered as a special class of one-layer NNs.

If the first-layer weights V are fixed and only the output-layer weights W are tuned, we have a one-layer NN. We may define $\phi(x) = \sigma(V^T x)$ so that the one-layer NN is described by

$$y = W^T \phi(x). \tag{1.1.7}$$

We shall see that it is important to take care in selecting the function $\phi(x)$ to obtain suitable performance from one-layer NNs. Note that $\phi(x)$ is a vector function $\varphi(x) : R^n \to R^L$.

Functional-link basis neural networks

More generality is gained if $\sigma(.)$ is not diagonal, but $\phi(x)$ is allowed to be a general function from R^n to R^L. This is called a *functional-link neural net* (*FLNN*) (Sadegh (1993)). FLNNs allow special structures of NNs to be created. The key to making NNs of special structures lies in the appropriate selection of the activation functions *and the thresholds*. Some special cases of FLNN are now considered.

Gaussian or radial basis function nets

An NN activation function often used is the Gaussian or radial basis function (RBF) (Sanner and Slotine (1991)), given when x is a scalar by

$$\sigma(x) = e^{-(x-\bar{x})^2/2p}, \tag{1.1.8}$$

where $\bar{x}$ is the mean and p the variance. Due to the fact that Gaussian functions are well understood in multivariable systems theory, the n-dimensional (nD) Gaussian function is easy to conceptualize.

In the nD case, the jth activation function can be written as

$$\sigma_j(x) = e^{-(x-\bar{x}_j)^T P_j^{-1}(x-\bar{x}_j)/2} \tag{1.1.9}$$

with $x, \bar{x}_j \in R^n$. Define the vector of activation functions as

$$\sigma(x) = \begin{bmatrix} \sigma_1(x) & \sigma_2(x) & \cdots & \sigma_L(x) \end{bmatrix}^T.$$

If the covariance matrix is diagonal so that $P_j = \text{diag}\{p_{jk}\}$, then (1.1.9) is *separable* and may be decomposed into components as

$$\sigma_j(x) = e^{-\sum_{k=1}^n (x_k-\bar{x}_{jk})^T (x_k-\bar{x}_{jk})/2p_{jk}} = \prod_{k=1}^n e^{-(x_k-\bar{x}_{jk})^2/2p_{jk}}, \tag{1.1.10}$$

where $x_k, \bar{x}_{jk}$ are the kth components of $x, \bar{x}_j$. Thus nD activation functions are the products of n scalar functions.

Note that (1.1.10) is of the form of the activation functions in (1.1.1) but with *more general activation functions*, as a different threshold is required for each different component of x at each hidden-layer neuron j. That is, the threshold at each hidden-layer neuron in Figure 1.1.1 is a *vector*. The offsets $\bar{x}_{jk}$ play the role of thresholds. The RBF variances p_{jk} and offsets $\bar{x}_{jk}$ are usually selected in designing the NN and then left fixed; only the output-layer weights W are generally tuned. Therefore, the RBF NN is a special case of the FLNN.

Figure 1.1.3 shows separable Gaussians for the case $x \in R^2$. In this figure, all the variances are identical and the offset vectors are selected to space the activation functions evenly over a grid. This is not possible using the scalar thresholds appearing in typical NNs. There are $L = 5 \times 5 = 25$ hidden-layer neurons for the NN in the figure, with 9 of the activation functions being Gaussian. Note that the boundary activation functions in the figure are selected to be one-sided Gaussians.

Random vector functional-link nets

If the number of hidden-layer neurons is large, it can be tedious and computationally prohibitive to select threshold vectors so that the activations uniformly cover a region in R^n. One faces at once the curse of dimensionality. One solution is to use vector thresholds, but to select these vectors *randomly*. In R^2 this corresponds to placing Gaussians at random over a specified region.

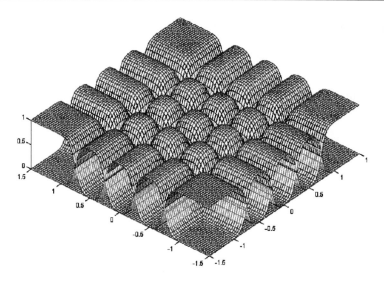

Figure 1.1.3. *Separable Gaussian activation functions for RBF NNs.*

It is also possible to use the standard NN

$$y = W^T \sigma(V^T x)$$

but simply select the weights V randomly (Igelnik and Pao (1995)). This selects randomly both the first-layer weights and the scalar hidden-layer thresholds. This means that the shifts and scaling of the activation functions are random. These weights are then fixed, and only W is tuned in the application.

Cerebellar model articulation controller nets

A cerebellar model articulation controller (CMAC) NN (Albus (1975)) has separable activation functions generally composed of splines. The activation functions of a two-dimensional (2D) CMAC consist of second-order splines (e.g., triangle functions) as shown in Figure 1.1.4, where $L = 5 \times 5 = 25$. The activation functions of CMAC NNs are called *receptive field functions* in analogy with the optical receptors of the eye. An advantage of CMAC NNs is that activation functions based on splines have *finite support* so that they may be efficiently evaluated. An additional computational advantage is provided by the fact that high-order splines may be recursively computed from lower order splines.

   CMAC NNs mimic the action of the cerebellar cortex in animals, since the receptor inputs are first collected into receptive fields using the activation functions; then the outputs of the activation functions are multiplied by weights W, which effectively perform some signal processing.

1.1.3   Universal approximation property of neural networks

NNs have many important properties including association, classification, and generalization (Haykin (1994)). An essential property for feedback control purposes is the *universal approximation property* (Cybenko (1989), Hornik et al. (1985), Park and Sandberg (1991)).

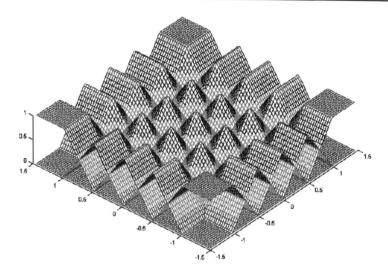

Figure 1.1.4. *Separable triangular activation functions for CMAC NNs.*

Approximation using two-layer neural networks

Let $f(x)$ be a general smooth function from R^n to R^m. Then it can be shown that, as long as x is restricted to a compact set S of R^n, there exist NN weights and thresholds such that one has

$$f(x) = W^T \sigma(V^T x) + \varepsilon \qquad (1.1.11)$$

for some number L of hidden-layer neurons. This holds for a large class of activation functions, including those in Figure 1.1.2. The functional reconstruction error ε can be made as small as desired. Generally, ε decreases as L increases. In fact, for any positive number ε_N, one can find weights and an L such that

$$\|\varepsilon\| < \varepsilon_N$$

for all $x \in S$. The unknown ideal weights required for good approximation are generally determined using some tuning procedure.

The selection of the required number L of hidden-layer neurons is an open problem for general NNs. However, consider, for instance, the separable 2D triangular activation functions in Figure 1.1.4. One can imagine a prescribed function $f(x) : R^2 \to R$ to be a surface above the x_1, x_2 plane. Then one must select output weights W that scale the activation functions to reach the surface. If one selects enough activation functions, one can achieve arbitrary accuracy in approximating $f(x)$ in this fashion. This involves selecting a grid with cell sizes small enough in the x_1, x_2 plane and then placing a shifted activation function at each node on the grid.

One can determine how small the cell size should be in the grid for a prescribed accuracy ε_N. This depends on the rate of change of $f(x)$. The details are given in Commuri (1996) or Sanner and Slotine (1991).

This approach allows one to select fixed first-layer weights and thresholds of the NN to achieve a given approximation accuracy for a class of functions $f(x)$. Then the output weights W required to approximate a specific given function $f(x)$ must be determined by some tuning procedure.

Approximation using one-layer nets

Though two-layer NNs generally enjoy the universal approximation property, the same is not true of one-layer NNs. For a one-layer NN,

$$y = W^T \phi(x); \tag{1.1.12}$$

to approximate functions, one must select the functions $\phi(x)$ to be a *basis* (Sadegh (1993)). That is, given a compact, simply connected function $S \subset R^n$, one must have the following conditions:

(1) a constant function can be expressed as (1.1.12) for a finite number L of hidden-layer neurons;

(2) the functional range of (1.1.12) is dense in the space of continuous functions from S to R^m for countable L.

Note that selecting a basis also means choosing L.

If $\phi(x)$ provides a basis, then for a general smooth function $f(x)$ from R^n to R^m, it can be shown that as long as x is restricted to a compact set S of R^n, there exist NN weights and thresholds W such that one has

$$f(x) = W^T \varphi(x) + \varepsilon. \tag{1.1.13}$$

The weights that offer suitable approximation accuracy may be determined by some tuning algorithm.

Selecting a basis

When using one-layer NNs, selecting a basis is a major problem in practical implementations. This problem does not arise with two-layer NNs, for through suitable tuning, the weights V can be modified to automatically generate a basis $\phi(x) = \sigma(V^T x)$ for the problem at hand. One possibility for selecting a basis is to use separable Gaussian or spline functions as in Figures 1.1.3 and 1.1.4. It has also been shown (Igelnik and Pao (1995)) that if the first-layer weights and thresholds V in (1.1.3) are selected randomly, then the random vector functional link (RVFL) with activation functions $\phi(x) = \sigma(V^T x)$ provides a basis.

Linearity in the parameters

The one-layer NN have a large advantage in that they are *linear* in the tunable parameters (LIP). This makes feedback control design far easier than using multilayer NNs, which are nonlinear in the tunable parameters (NLIP). (Note that the weights V appear in the argument of the nonlinear functions $\sigma(.)$.) However, there is a price to pay, for Barron (1993) has shown that for all LIP networks, there is a fundamental lower bound on the approximation accuracy. In fact, the approximation error ε is bounded below by terms on the order of $1/L^{2/n}$. This means that the error decreases as L increases, but that as the input dimension n becomes large, increasing L has a diminishing ability to decrease ε.

1.2   Fuzzy logic systems

In this section we will provide a background for FL systems (Kosko (1997), Passino and Yurkovich (1998), Ying (2000), Wang (1997)), focusing on properties needed for feedback control design. There are many ways to bring together NNs and FL systems (Kosko (1992)), including architectures having both NN and FL components, using FL systems to initialize NN weights or NNs to adapt FL membership functions, and so on. Numerous papers have appeared on these relationships. However, one point of view is to consider FL systems as a special class of *structured NNs*. RBF and CMAC NNs introduce structure to the NN by using *separability* of the nD activation functions into products of one-dimensional (1D) functions. FL networks introduce additional structure by formalizing the selection of the NN weights and thresholds through relating these 1D functions to concepts in human linguistics.

A standard FL system is shown in Figure 1.2.1. It has three basic stages. *Fuzzification* is accomplished using the *membership functions* (MFs), which assign high-level linguistic meaning to values of the inputs x_k. The *inference mechanism* is usually specified using a rule-base and associates the fuzzified inputs to high-level descriptions of the outputs. Finally, using some *defuzzification* scheme, one obtains the so-called crisp values of the outputs y_i, which are numerical values at specified points in time, i.e., simply standard time functions.

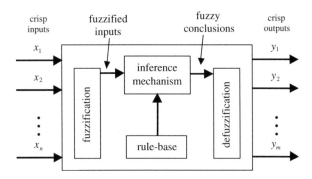

Figure 1.2.1. *FL system.*

We shall now see that the FL system of Figure 1.2.1 is closely related to the NN of Figure 1.1.1. In fact, it can be shown that FL systems using product inferencing and weighted defuzzification are equivalent to special sorts of NNs with suitably chosen separable activation functions, variances, and offsets.

1.2.1   Membership functions and defuzzification

Refer at this point to (1.1.1). In FL systems a scalar function $\sigma(.)$ is selected, often the triangle function. Then for $x \in R^n$ with components x_k, the scalar function

$$X_j^k(x_k) = \sigma(v_{jk}x_k + \theta_{v_{jk}}) \tag{1.2.1}$$

is the jth MF along component x_k, shifted by $\theta_{v_{jk}}$ and scaled by v_{jk}. See Figure 1.2.2. Using product inferencing, the nD MFs are composed using multiplication of the scalar MFs so that

$$X_{j1,j2,...,jn}(x) = X_{j1}^1(x_1)X_{j2}^2(x_2)\cdots X_{jn}^n(x_n) \tag{1.2.2}$$

for a specified index set $\{j1, j2, \ldots, jn\}$. Thus in FL nets, the nD MFs are by construction separable in terms of products of 1D functions, as in (1.1.10).

FL systems are very closely related to the RBF and CMAC NNs. In fact, the RBF NN in Figure 1.1.3 is equivalent to a 2D fuzzy system with Gaussian MFs along x_1 and x_2, while the CMAC NN in Figure 1.1.4 is equivalent to a 2D FL system with triangle MFs

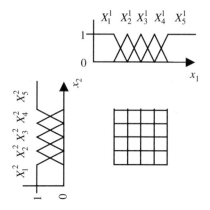

Figure 1.2.2. *FL MFs for the 2D case.*

(i.e., multiply the MFs in Figure 1.2.2 to get Figure 1.1.4). Note that FL systems have more general thresholds than (1.1.1) since, as for RBF and CMAC NNs, the thresholds at each hidden-layer neuron must be n-vectors.

With $y \in R^m$ having components y_i, using centroid defuzzification, the ith output component is given as

$$y_i = \frac{\sum w_{i,j1,j2,\ldots,jn} X_{j1,j2,\ldots,jn}(x)}{\sum X_{j1,j2,\ldots,jn}(x)}, \tag{1.2.3}$$

where the $w_{i,j1,j2,\ldots,jn}$ are known in FL systems as the *control representative values*. The associations of the output index set $\{i\}$ to the MF index sets $\{j1, j2, \ldots, jn\}$ and the values of $w_{i,j1,j2,\ldots,jn}$ are given by a set of L "rules." The summation is performed over all the rules. This equation is of the form (1.1.3) with appropriate definitions of the activation functions and the weights W and V.

1.2.2   Universal approximation property and tuning of fuzzy logic networks

It has been shown by several authors (Wang and Mendel (1992)) that FL nets possess the universal approximation property. That is, there exist $v_{jk}, \theta_{v_{jk}}, w_{i,j1,j2,\ldots,jn}$ so that (1.2.3) can approximate any sufficiently smooth scalar function $f(x)$ to within any desired accuracy. The normalized functions

$$\bar{X}_{j1,j2,\ldots,jn}(x) = \frac{X_{j1,j2,\ldots,jn}(x)}{\sum X_{j1,j2,\ldots,jn}(x)} \tag{1.2.4}$$

are called the *fuzzy basis functions* and correspond to the hidden-layer output functions $\sigma(.)$ in (1.1.1).

In *adaptive FL systems*, one may tune the parameters using the NN backpropagation algorithm. One may adapt the control representative values W and/or the MF parameters V. If the first-layer weights and thresholds V are fixed so that the MFs are not tuned, and only W is tuned, then the FL net is linear in the parameters; these FL systems are therefore FLNNs as in (1.1.7) and the MFs must be chosen as a basis on some compact set. This may be accomplished as discussed in the "Selecting a basis" section above for NNs.

If both W and V are adapted, the FL system is NLIP and possesses the universal approximation property (1.1.11). Then for tuning V, one requires MFs that are differentiable. The Jacobians $\sigma'(.)$ required for backprop tuning in the NLIP case using, e.g., the Gaussian MFs, are easy to compute.

Chapter 2

Background on Dynamical Systems and Industrial Actuators

In this chapter, we shall provide a brief background on dynamical systems, covering the topics needed for the design of intelligent control systems for industrial actuators. Included are dynamics, stability, and Lyapunov proof techniques. More details may be found in Lewis, Jagannathan, and Yesildirek (1999). We also discuss some nonlinear effects of industrial actuators, studying friction, deadzone, and backlash. This provides the material required to begin a study of neural and fuzzy control techniques in improving the performance of industrial actuators (Selmic (2000), Campos Portillo (2000)), which we do in subsequent chapters.

2.1 Dynamical systems

Many systems in nature, including biological systems, are *dynamical* in the sense that they are acted upon by external inputs, have internal memory, and behave in certain ways that can be captured by the notion of the development of activities through time. The notion of *system* was formalized in the early 1900s by Whitehead (1953) and von Bertalanffy (1968). A system is viewed here as an entity distinct from its environment, whose interactions with the environment can be characterized through *input* and *output* signals. An intuitive feel for dynamical systems is provided by Luenberger (1979), which has many excellent examples.

2.1.1 Continuous and discrete dynamical systems

Continuous-time systems

A very general class of continuous-time systems can be described by the nonlinear ordinary differential equation in *state-space form*

$$\dot{x} = F(x, u),$$
$$y = H(x, u), \qquad (2.1.1)$$

where $x(t) \in R^n$ is the internal state vector, $u(t) \in R^m$ is the control input, and $y(t) \in R^p$ is the measured system output. An overdot represents differentiation with respect to time t. This state equation can describe a variety of dynamical behaviors, including mechanical and electrical systems, earth atmosphere dynamics, planetary orbital dynamics, aircraft systems, population growth dynamics, and chaotic behavior.

A special case of this system is

$$\dot{x} = f(x) + g(x)u,$$
$$y = h(x),$$

(2.1.2)

which is linear or affine in the control input.

Letting $x = [x_2 \quad x_2 \quad \cdots \quad x_n]^T$, a special important form of nonlinear continuous-time dynamics is given by the class of systems in *Brunovsky canonical form*,

$$x_1 = x_2,$$
$$\dot{x}_2 = x_3,$$
$$\vdots$$
$$\dot{x}_n = f(x) + g(x)u,$$
$$y = h(x).$$

(2.1.3)

As seen from Figure 2.1.1, this is a chain or cascade of integrators $1/s$; each integrator stores information and requires an initial condition. The internal state $x(t)$ can be viewed as the initial information required to specify a unique solution of the differential equation.

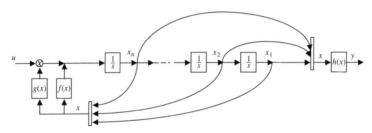

Figure 2.1.1. *Continuous-time single-input Brunovsky form.*

Another special form of the continuous-time state equation is given by the *linear-time invariant (LTI) system*

$$\dot{x} = Ax + Bu,$$
$$y = Cx + Du.$$

(2.1.4)

Many naturally occurring systems are linear, and feedback control design for LTI systems is far easier than for nonlinear systems.

Discrete-time systems

If the time index is an integer k instead of a real number t, the system is said to be of *discrete time*. A very general class of discrete-time systems can be described by the nonlinear difference equation in discrete state-space form

$$x(k + 1) = F(x(k), u(k)),$$
$$y(k) = H(x(k), u(k)),$$

(2.1.5)

where $x(k) \in R^n$ is the internal state vector, $u(k) \in R^m$ is the control input, and $y(k) \in R^p$ is the measured system output.

Today, controllers are generally implemented in digital form so that a discrete-time description of the controller is needed. Discrete-time equations have great advantages in computer simulation and implementation, since integration of continuous-time dynamics requires an integration routine such as Runge–Kutta, while discrete-time dynamics may be programmed using a simple iterative algorithm.

The discrete-time equations may either be derived directly from an analysis of the dynamical process being studied or be *sampled* or discretized versions of continuous-time dynamics in the form (2.1.1). Sampling of linear systems has been well understood since the work of Ragazzini, Franklin, and others in the 1950s, with many design techniques available. However, sampling of nonlinear systems is not an easy topic. In fact, the exact discretization of nonlinear continuous dynamics is based on the Lie derivatives and leads to an infinite series representation. Various approximate discretization techniques use truncated versions of the exact series.

With $x_i(k)$ the components of the n-vector $x(k)$, a special form of nonlinear dynamics is given by the class of systems in *discrete Brunovsky canonical form*,

$$x_1(k + 1) = x_2(k),$$
$$x_2(k + 1) = x_3(k),$$
$$\vdots \qquad\qquad (2.1.6)$$
$$x_n(k + 1) = f(x(k)) + g(x(k))u(k),$$
$$y(k) = h(x(k)).$$

As seen from Figure 2.1.2, this is a chain or cascade of unit delay elements z^{-1}, i.e., a *shift register*. Each delay element stores information and requires an initial condition.

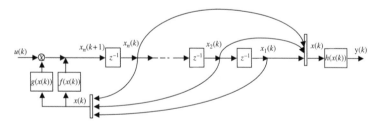

Figure 2.1.2. *Discrete-time single-input Brunovsky form.*

The linear-time invariant form of discrete-time dynamics is given by

$$x(k + 1) = Ax(k) + Bu(k),$$
$$y = Cx(k) + Du(k). \qquad (2.1.7)$$

2.1.2   Mathematics and system properties

Mathematical background

We denote any suitable norm by $\|.\|$ and may specifically denote the p-norm by $\|.\|_p$ when required. For any vector $x \in R^n$ with components x_i, recall that the 1-norm is the sum of the absolute values of x_i, the 2-norm is the square root of the sum of the squares of x_i, and the ∞-norm is the maximum absolute value of the x_i. Generally,

$$\|x\|_p = \left(\sum_{i=1}^{n} |x_i|^p\right)^{1/p}.$$

Recall that all vector norms are equivalent in the sense that for any vector x, there exist constants a_1, a_2 such that $a_1\|x\|_p \leq \|x\|_q \leq a_2\|x\|_p$ for any norms p, q. This means that any norm can be used to show boundedness of signals or convergence.

Given any vector p-norm, the induced matrix norm is defined by

$$\|A\|_p \equiv \sup_{x\neq 0} \frac{\|Ax\|_p}{\|x\|_p}.$$

Recall that $\|A\|_1$ is equal to the maximum absolute column sum, $\|A\|_\infty$ is the maximum absolute row sum, and $\|A\|_2$ is the maximum singular value of A. The induced matrix norm satisfies

$$\|Ax\|_p \leq \|A\|_p\|x\|_p.$$

For any two matrices A, B, one also has

$$\|AB\|_p \leq \|A\|_p\|B\|_p.$$

A special matrix norm is the Frobenius norm defined by

$$\|A\|_F^2 = \text{trace}(A^T A) = \sum a_{ij}^2$$

with a_{ij} the elements of A. The Frobenius norm is not an induced norm for any vector norm, but it is compatible with the 2-norm in the sense that

$$\|Ax\|_2 \leq \|A\|_F\|x\|_2.$$

The Frobenius inner product is $\langle A, B\rangle_F = A^T B$ for any compatibly dimensioned matrix B. According to the Schwarz inequality, one has

$$|\langle A, B\rangle_F| \leq \|A\|_F\|B\|_F.$$

Unless otherwise stated, all matrix norms used in this book are Frobenius norms and all vector norms will be 2-norms.

A matrix Q is positive definite if $x^T Qx > 0$ for all $x \neq 0$, positive semidefinite if $x^T Qx \geq 0$ for all $x \neq 0$, negative definite if $x^T Qx < 0$ for all $x \neq 0$, and negative semidefinite if $x^T Qx \leq 0$ for all $x \neq 0$. These are, respectively, denoted by $Q > 0$, $Q \geq 0$, $Q < 0$, and $Q \leq 0$. If Q is symmetric, then it is positive definite if all its eigenvalues are positive, positive semidefinite if all eigenvalues are nonnegative, negative definite if all eigenvalues are negative, and negative semidefinite if all eigenvalues are nonpositive.

A norm for vector x is also defined by $x^T Qx$, where Q is any symmetric positive definite matrix. This Q-norm satisfies the Sylvester inequality

$$\sigma_{\min}(Q)\|x\|^2 \leq x^T Qx \leq \sigma_{\max}(Q)\|x\|^2$$

with $\sigma_{\min}, \sigma_{\max}$ the minimum and maximum singular value, respectively. This also means that

$$-x^T Qx \leq \sigma_{\min}(Q)\|x\|^2.$$

Stability and boundedness of systems

Consider the dynamical system

$$\dot{x} = f(x, t) \tag{2.1.8}$$

with $x(t) \in R^n$. A state x_e is an equilibrium point of the system if $f(x_e t) = 0$, $t \geq t_0$, with t_0 the initial time. For linear systems, the only possible equilibrium point is $x_e = 0$. For nonlinear systems, not only may $x_e \neq 0$, but there may be equilibrium trajectories (e.g., limit cycles), not only equilibrium points.

An equilibrium point x_e is locally asymptotically stable (AS) at t_0 if there exists a compact set $S \subset R^n$ such that for every initial condition $x(t_0) \in S$, one has $\|x(t) - x_e\| \to 0$ as $t \to \infty$. That is, the state $x(t)$ converges to x_e. If $S = R^n$ so that $x(t)$ converges to x_e for all initial conditions, then x_e is said to be globally asymptotically stable (GAS) at t_0. If this holds for all t_0, the stability is said to be uniform (UAS, GUAS).

Asymptotic stability is a strong property that is very difficult to achieve in practical dynamical systems. Therefore, we have a few milder definitions that are more useful.

An equilibrium point x_e is stable in the sense of Lyapunov (SISL) at t_0 if for every $\varepsilon > 0$ there exists a $\delta(\varepsilon, t_0) > 0$ such that $\|x(t_0) - x_e\| < \delta(\varepsilon, t_0)$ implies that $\|x(t) - x_e\| < \varepsilon$ for $t \geq t_0$. The stability is said to be uniform (e.g., uniformly SISL) if $\delta(.)$ is independent of t_0. That is, the system is SISL for all t_0.

Note that SISL requires the ability to keep the state $x(t)$ arbitrarily close to x_e by starting sufficiently close to it. This is still too strong a requirement for practical systems stability analysis in the presence of disturbances. Therefore, we are going to settle on the next definition.

This definition is related to Figure 2.1.3. The equilibrium point x_e is uniformly ultimately bounded (UUB) if there exists a compact set $S \subset R^n$ such that for every $x(t_0) \in S$, there exists a bound B and a time $T(B, x(t_0))$ such that $\|x(t) - x_e\| < B$ for all $t \geq t_0 + T$. The intent here is to capture the notion that for all initial states in the compact set S, the system trajectory eventually reaches after a lapsed time of T, a bounded neighborhood of x_e.

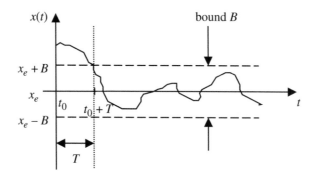

Figure 2.1.3. *Illustration of uniform ultimate boundedness.*

The difference between uniform ultimate boundedness and SISL is that the bound B cannot be made arbitrarily small by starting closer to x_e. In practical systems, the bound B depends on disturbances and other factors. A good controller design should, however, guarantee that B is small enough. The term *uniform* indicates that B does not depend on t_0. The term *ultimate* indicates that boundedness holds after the lapse of a certain time T. If $S = R^n$, the system is said to be globally UUB (GUUB).

Note that if the function $f(.)$ in (2.1.8) does not depend explicitly on the time t, the stability or boundedness is always uniform if it holds. Nonuniformity is only a problem with time-dependent systems.

2.1.3 Lyapunov stability analysis

Industrial systems are generally nonlinear, a problem that is exacerbated by the presence of actuator nonlinearities such as friction, deadzones, and backlash. Strict modern-day tolerances in semiconductor very large-scale integration (VLSI) manufacturing, high-precision machining of parts for aerospace systems, and other fields pose severe requirements on positioning speed and accuracy of mechanical industrial systems. These requirements cannot be met without confronting the full nonlinearities of the actuators. The design of control systems for nonlinear systems is not easy. In this section we discuss some techniques for nonlinear control system design.

Lyapunov stability theorem

The autonomous (time-invariant) dynamical system

$$\dot{x} = f(x), \tag{2.1.9}$$

$x(t) \in R^n$, could represent a closed-loop system after the controller has been designed. We shall show here how to examine the stability properties using a generalized energy approach. An isolated equilibrium point x_e can always be brought to the origin by redefinition of coordinates; therefore, let us assume without loss of generality that the origin is an equilibrium point.

A function $L(x) : R^n \to R$ with continuous partial derivatives is said to be a *Lyapunov function* for the system (2.1.9) if for some compact set $S \subset R^n$, one has that

$L(x)$ is positive definite, e.g., $L(x) > 0$ for all nonzero x,
$\dot{L}(x)$ is negative semidefinite, e.g., $\dot{L}(x) \leq 0$ for all nonzero x,

where $\dot{L}(x)$ is evaluated along the trajectories of (2.1.9). That is,

$$\dot{L}(x) = \frac{\partial L}{\partial x}\dot{x} = \frac{\partial L}{\partial x}f(x).$$

The Lyapunov function is a generalized energy function. The next main result shows that it can be used to analyze stability for the system (2.1.9). We shall see how to use this result to design control systems in this book.

Theorem 2.1.1 (Lyapunov stability). *If there exists a Lyapunov function for system* (2.1.9), *then the equilibrium point is SISL.*

The next theorem provides a stronger result if the Lyapunov derivative is negative definite.

Theorem 2.1.2 (Asymptotic stability). *If there exists a Lyapunov function for system* (2.1.9) *with the strengthened condition* $\dot{L}(x) < 0$ *for all nonzero* $x(t)$, *then the equilibrium point is AS.*

Example 2.1.1 (Asymptotic stability).

(a) *Local stability*. Consider the system

$$\dot{x}_1 = x_1 x_2^2 + x_1(x_1^2 + x_2^2 - 3),$$
$$\dot{x}_2 = -x_1^2 x_2 + x_2(x_1^2 + x_2^2 - 3).$$

Stability may often be examined for nonlinear systems by selecting quadratic Lyapunov functions like the candidate

$$L(x) = \frac{1}{2}(x_1^2 + x_2^2),$$

which is a direct generalization of an energy function and has the derivative

$$\dot{L}(x) = x_1 \dot{x}_1 + x_2 \dot{x}_2.$$

Evaluating this along the system trajectories simply involves substituting the state derivatives from the dynamics to obtain

$$\dot{L}(x) = -(x_1^2 + x_2^2)(3 - x_1^2 - x_2^2),$$

which is negative as long as

$$\|x\|^2 = x_1^2 + x_2^2 < 3.$$

Therefore, $L(x)$ serves as a (local) Lyapunov function for the system, which is locally AS. The system is said to have a domain of attraction with radius of 3.

(b) *Global stability*. Now consider the system

$$\dot{x}_1 = x_1 x_2^2 - x_1(x_1^2 + x_2^2),$$
$$\dot{x}_2 = -x_1^2 x_2 - x_2(x_1^2 + x_2^2).$$

Again, selecting the Lyapunov function candidate

$$L(x) = \frac{1}{2}(x_1^2 + x_2^2)$$

yields

$$\dot{L}(x) = -(x_1^2 + x_2^2)^2,$$

which is negative. The function $L(x)$ is therefore a Lyapunov function and the system is GAS.

Example 2.1.2 (Lyapunov stability). For the system

$$\dot{x}_1 = x_1 x_2^2 - x_1,$$
$$\dot{x}_2 = -x_1^2 x_2,$$

select the Lyapunov function candidate

$$L(x) = \frac{1}{2}(x_1^2 + x_2^2),$$

which has

$$\dot{L}(x) = -x_1^2.$$

This is only negative semidefinite since it can be zero when $x_2 \neq 0$. Therefore, the system is only SISL; that is, the internal state is bounded but does not go to zero with time.

Lyapunov extensions and bounded stability

In some situations, these two theorems provide too little information and yet turn out to be asking too much from industrial and defense systems that have disturbances and actuator deficiencies. Some extensions of the Lyapunov theory lead to more practical results that can be used in actual industrial control systems. The first result is based on Barbalat's lemma.

Given a subset $S \subset R^n$, a function $f(x) : S \to R^m$ is *continuous* on S if for every $x_0 \in S$ and for every $\varepsilon > 0$ there exists a $\delta(e, x_0) > 0$ such that $\|x - x_0\| < \delta(e, x_0)$ implies that $\|f(x) - f(x_0)\| < \varepsilon$.

If $\delta(.)$ is independent of x_0, then $f(x)$ is *uniformly continuous*. Uniform continuity is often difficult to test for. However, if $f(x)$ is continuous and its derivative $\partial f / \partial x$ is bounded, then $f(x)$ is uniformly continuous. This test is only sufficient but is easier to use.

Theorem 2.1.3 (Barbalat's lemma, Lyapunov extension). *Let $L(x)$ be a Lyapunov function. If $\dot{L}(x)$ is uniformly continuous, then $\dot{L}(x) \to 0$ as $t \to \infty$.*

This theorem allows us to get more information about the system behavior than is provided by Theorem 2.1.1. It says that if $\dot{L}(x)$ has an extra property, then one can conclude that $\dot{L}(x) \to 0$. This can often allow us to say that certain states in the system are not simply bounded but actually go to zero.

To verify the uniform continuity of $\dot{L}(x)$, it is only necessary to check whether $\ddot{L}(x)$ is bounded.

Example 2.1.3 (Asymptotic stability of state components using Barbalat's extension). In Example 2.1.2, we considered the system

$$\dot{x}_1 = x_1 x_2^2 - x_1,$$
$$\dot{x}_2 = -x_1^2 x_2$$

with the Lyapunov function candidate

$$L(x) = \frac{1}{2}(x_1^2 + x_2^2),$$

which has

$$\dot{L}(x) = -x_1^2.$$

Since this is negative semidefinite, we were able to conclude that the system is SISL.

Using Barbalat's lemma, we can extract more information about the stability from this Lyapunov function. In fact, note that

$$\ddot{L}(x) = -2x_1 \dot{x}_1 = -2x_1^2 x_2^2 + 2x_1^2.$$

The Lyapunov analysis showed SISL, which means that $\|x\|$ is bounded. This in turn shows that $\ddot{L}(x)$ is bounded, hence that $\dot{L}(x)$ is uniformly continuous. Therefore, $\dot{L}(x)$ actually goes to zero with time. This, finally, shows that $x_1(t) \to 0$. Thus Barbalat's extension has allowed us to prove that certain state components are AS.

The next result provides a milder form of stability than SISL that is more useful for controller design in practical systems, which usually have the form

$$\dot{x} = f(x) + d(t) \qquad (2.1.10)$$

with $d(t)$ an unknown but bounded disturbance. The theorem tells when uniform ultimate boundedness is guaranteed.

Theorem 2.1.4 (Uniform ultimate boundedness by Lyapunov extension). *Suppose that for system (2.1.10) there exists a function $L(x)$ with continuous partial derivatives such that for x in a compact set $S \subset R^n$,*

$$L(x) \text{ is positive definite,}$$
$$\dot{L}(x) \text{ is negative definite for } \|x\| > R$$

for some $R > 0$ such that the ball of radius R is contained in S. Then the system is UUB and, moreover, the norm of the state is bounded to within a neighborhood of the ball of radius R.

Example 2.1.4 (UUB Lyapunov extension). Consider the system

$$\dot{x}_1 = x_1 x_2^2 - x_1(x_1^2 + x_2^2 - 3),$$
$$\dot{x}_2 = -x_1^2 x_2 - x_2(x_1^2 + x_2^2 - 3)$$

and select the Lyapunov function candidate

$$L(x) = \frac{1}{2}(x_1^2 + x_2^2).$$

This has the derivative

$$\dot{L}(x) = -(x_1^2 + x_2^2)(x_1^2 + x_2^2 - 3),$$

which is negative as long as

$$\|x\|^2 = x_1^2 + x_2^2 > 3.$$

Standard Lyapunov techniques fail here since $\dot{L}(x)$ is not even negative semidefinite. However, the UUB extension shows that the system is UUB with the state restricted to the neighborhood of the ball with radius $\|x\| = 3$.

2.2 Industrial actuators

Industrial processes such as CNC machines, robots, VLSI positioning tables, and so on are moved by *actuators*. An actuator is a device that provides the motive power to the process by mechanically driving it. There are many classifications of actuators. Those that directly operate a process are called *process actuators*. Joint motors in robotic manipulators are process actuators, for example. In process control applications, actuators are also often used to operate controller components, such as servovalves. Actuators in this category are called *control actuators*. Most actuators used in control applications are continuous-drive devices. Examples are direct current (DC) torque motors, induction motors, hydraulic and pneumatic motors, and piston–cylinder drives. There are incremental-drive actuators like stepper motors; these actuators can be treated as digital actuators.

Mechanical parts and elements are unavoidable in all actuator devices. Inaccuracies of mechanical components and the nature of physical laws mean that all these actuator devices are nonlinear. If the input–output relations of the device are nonlinear algebraic equations, this represents a *static nonlinearity*. On the other hand, if the input–output relations are nonlinear differential equations, it represents a *dynamic nonlinearity*. Examples of nonlinearities include friction, deadzone, saturation (all static), and backlash and hysteresis (both dynamic).

A general class of industrial processes has the structure of a dynamical system preceded by the nonlinearities of the actuator. Problems in controlling these processes are particularly exacerbated when the required accuracy is high, as in micropositioning devices. Due to the nonanalytic nature of the actuator nonlinearities and the fact that their exact nonlinear functions are unknown, such processes present a challenge for the control design engineer.

2.3   Friction

Friction is a natural resistance to relative motion between two contacting bodies and is essential for the operation of common mechanical systems (think of wheels, clutches, etc). But in most industrial processes it also represents a problem, since it is difficult to model and troublesome to deal with in control system design. Manufacturers of components for precision control systems take efforts to minimize friction, and this represents a significant increase in costs. However, notwithstanding efforts at reducing friction, its problems remain and it is necessary to contend with them in precision control systems. The possible undesirable effects of friction include "hangoff" and limit cycling. Hangoff prevents the steady-state error from becoming zero with a step command input (this can be interpreted as a DC limit cycle). Limit cycling is the behavior in which the steady-state error oscillates about zero.

Friction is a complicated nonlinear phenomenon in which a force is produced that tends to oppose the motion in a mechanical system. Motion between two contacting bodies causes the dissipation of energy in the system. The physical mechanism for the dissipation depends on the materials of the rubbing surfaces, their finish, the lubrication applied, and other factors, many of which are not yet fully understood.

The concern of the control engineer is not reducing friction but dealing with friction that cannot be reduced. To compensate for friction, it is necessary to understand and have a model of the friction process. Many researchers have studied friction modeling. Extensive work can be found in Armstrong-Hélouvry, Dupont, and Canudas de Wit (1994).

Static friction models

The classic model of friction force that is proportional to load, opposes the motion, and is independent of contact area was known to Leonardo da Vinci. Da Vinci's friction model was rediscovered by Coulomb and is widely used today as the simplest friction model, described by

$$F(v) = a\,\mathrm{sgn}(v), \qquad (2.3.1)$$

where v is the relative velocity and $F(v)$ is the corresponding force or torque. The parameter a is generally taken as a constant for simplicity. Coulomb friction is shown in Figure 2.3.1(a). A more detailed friction model is shown in Figure 2.3.1(b), which includes *viscous friction*, a term proportional to the velocity.

Physical experiments have shown that in many cases the force required to initiate relative motion is larger than the force that opposes the motion once it starts. This effect is known as *static friction* or *stiction*. Modeling stiction effects is accomplished by use of a nonlinearity of the form shown in Figure 2.3.1(c).

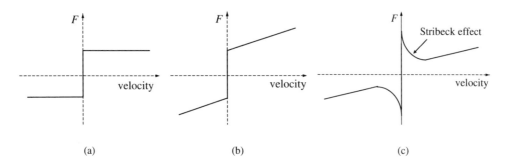

Figure 2.3.1. *Friction models:* (a) *Coulomb friction.* (b) *Coulomb and viscous friction.* (c) *Complete friction model.*

An empirical formula sometimes used for expressing the dependence of the friction force upon velocity is

$$F(v) = (a - be^{-c|v|} + d|v|)\,\text{sgn}(v), \tag{2.3.2}$$

in which the parameters a, b, c, and d are chosen to impart the desired shape to the friction function. A complete model for friction suitable for industrial controller design is given in Canudas de Wit et al. (1991) as

$$F(v) = [\alpha_0 + \alpha_1 e^{-\beta_1|v|} + \alpha_2(1 - e^{-\beta_2|v|})]\,\text{sgn}(v), \tag{2.3.3}$$

where Coulomb friction is given by α_0 (Nm), static friction is $(\alpha_0 + \alpha_1)$ (Nm), and α_2 (Nm-sec/rad) represents the viscous friction model. The effect whereby for small v the friction force is decreasing with velocity is called *negative viscous friction* or the *Stribeck effect*. The Stribeck effect is modeled with an exponential second term in the model (2.3.3). This friction model captures all the effects shown in Figure 2.3.1(c).

Dynamic friction models

Though friction is usually modeled as a static discontinuous map between velocity and friction torque, which depends on the velocity's sign, there are several interesting properties observed in systems with friction that cannot be explained only by static models. This is basically due to the fact that the friction does not have an instantaneous response to a change in velocity (i.e., it has internal dynamics). Examples of these dynamic properties are

- *stick-slip motion*, which consists of limit cycle oscillation at low velocities, caused by the fact that friction is larger at rest than during motion;

- *presliding displacement*, which means that friction behaves like a spring when the applied force is less than the static friction break-away force;

- *frictional lag*, which means that there is a hysteresis in the relationship between friction and velocity.

All these static and dynamic characteristics of friction were captured by the dynamical model and analytical model proposed in Canudas de Wit et al. (1995). This model is called

LuGre (the Lund–Grenoble model). The LuGre model is given by

$$\frac{dz}{dt} = \dot{q} - \frac{\sigma_0}{g(\dot{q})} z |\dot{q}|,$$

$$g(\dot{q}) = \alpha_0 + \alpha_1 e^{-\left(\frac{\dot{q}}{v_0}\right)^2}, \qquad (2.3.4)$$

$$F = \sigma_0 z + \sigma_1 \frac{dz}{dt} + \alpha_2 \dot{q},$$

where $\dot{q}$ is the angular velocity and F is the friction force. The first of these equations represents the dynamics of the friction internal state z, which describes the average relative deflection of the contact surfaces during the stiction phases. This state is not measurable. The function $g(\dot{q})$ describes the *steady-state* part of the model or constant velocity motions: v_0 is the Stribeck velocity, $(\alpha_0 + \alpha_1)$ is the static friction, and α_0 is the Coulomb friction. Thus the complete friction model is characterized by four static parameters, α_0, α_1, α_2, and v_0, and two dynamic parameters, σ_0 and σ_1. The parameter σ_0 can be understood as a stiffness coefficient of the microscopic deformations of z during the presliding displacement, while σ_1 is a damping coefficient associated with dz/dt.

2.4   Deadzone

Deadzone (Tao and Kokotović (1996)) is a static nonlinearity that describes the insensitivity of the system to small signals. Although there are some open-loop applications where the deadzone characteristic is highly desirable, in most closed-loop applications, deadzone has undesirable effects on the feedback loop dynamics and control system performance. It represents a "loss of information" when the signal falls into the deadband and can cause limit cycles, tracking errors, and so forth.

Deadzone has the static input–output relationship shown in Figure 2.4.1. A mathematical model is given by

$$\tau = D(u) = \begin{cases} m_-(u + d_-), & u \leq -d_-, \\ 0, & -d_- < u < d_+, \\ m_+(u - d_+), & u \geq d_+. \end{cases} \qquad (2.4.1)$$

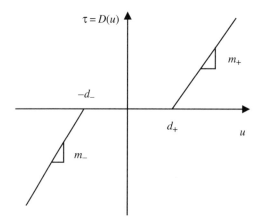

Figure 2.4.1. *Deadzone nonlinearity.*

One can see that there is no output as long as the input signal is in the deadband defined by $-d_- < u < d_+$. When the signal falls into this deadband, the output signal is zero, and one loses information about the input signal. Once the output appears, the slope between input and output stays constant. Note that (2.4.1) represents a nonsymmetric deadzone model since the slope on the left side and right side of the deadzone is not the same.

Deadzone characteristic (2.4.1) can be parametrized by the four constants d_-, m_-, d_+, and m_+. In practical motion control systems, these parameters are unknown, and compensation of such nonlinearities is difficult. Deadzones usually appear at the input of the actuator systems, as in the DC motor, but there are also output deadzones, where the nonlinearities appear at the output of the system.

A deadzone is usually caused by friction, which can vary with temperature and wear. Also, these nonlinearities may appear in mass-produced components, such as valves and gears, which can vary from one component to another.

An example of deadzone caused by friction given in Tao and Kokotović (1996) is shown in Figure 2.4.2. The input to the motor is motor torque T_m, the transfer function in the forward path is a first-order system with time constant τ. There is a Coulomb friction in the feedback path. If the time constant τ is negligible, the low-frequency approximation of the feedback loop is given by the deadzone characteristic shown in Figure 2.4.1. Note that the friction torque characteristic is responsible for the break points d_+ and d_-, while the feedforward gain m determines the slope of the deadzone function.

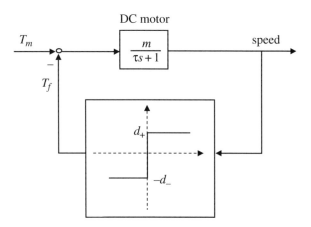

Figure 2.4.2. *Deadzone caused by friction in a DC motor.*

The deadzone may be written as

$$\tau = D_d(u) = u - \mathrm{sat}_d(u), \tag{2.4.2}$$

where the nonsymmetric saturation function is defined as

$$\mathrm{sat}_d(u) = \begin{cases} -d_-, & u < -d_-, \\ u, & -d_- \leq u < d_+, \\ d_+, & d_+ \leq u. \end{cases} \tag{2.4.3}$$

This decomposition, shown in Figure 2.4.3, represents a feedforward path plus an unknown parallel path and is extremely useful for control design.

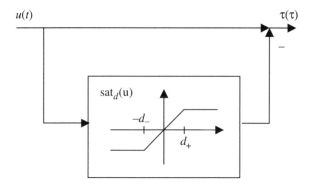

Figure 2.4.3. *Decomposition of deadzone into feedforward path plus unknown path.*

2.5 Backlash

The space between teeth on a mechanical gearing system must be made larger than the gear teeth width as measured on the pitch circle. If this were not the case, the gears could not mesh without jamming. The difference between toothspace and tooth width is known as backlash. Figure 2.5.1 shows the backlash present between two meshing spur gears. Any amount of backlash greater than the minimum amount necessary to ensure satisfactory meshing of gears can result in instability in dynamic situations as well as in position errors in gear trains. In fact, there are many applications such as instrument differential gear trains and servomechanisms that require the complete elimination of backlash in order to function properly.

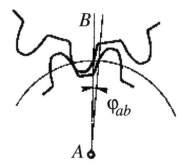

Figure 2.5.1. *Backlash in a gear system.*

Backlash results in a delay in the system motion. One can see that when the driving gear changes its direction, the driven gear follows only after some delay. A model of backlash in mechanical systems is shown in Figure 2.5.2.

A standard mathematical model (Tao and Kokotović (1996)) is given by

$$\dot{\tau} = B(\tau, u, \dot{u}) = \begin{cases} m\dot{u} & \text{if } \dot{u} > 0 \text{ and } \tau = mu - md_+, \\ & \text{if } \dot{u} < 0 \text{ and } \tau = mu - md_-, \\ 0 & \text{otherwise.} \end{cases} \qquad (2.5.1)$$

One can see that backlash is a first-order velocity-driven dynamical system with inputs u

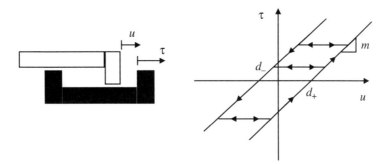

Figure 2.5.2. *Backlash nonlinearity.*

and $\dot{u}$ and state τ. It contains its own dynamics; therefore, its compensation requires the use of a dynamic compensator.

Whenever the driving motion $u(t)$ changes its direction, the resultant motion $\tau(t)$ is delayed from the motion of $u(t)$. The objective of a backlash compensator is to make this delay as small as possible.

Chapter 3

Neurocontrol of Systems with Friction

Here we develop a technique for control of industrial processes with friction using NNs. The NN appears in a feedback control loop and significantly improves the performance of the process by compensating for the deleterious effects of friction. The NN weights will be tuned in such a fashion that the NN learns about the friction nonlinearity on line. No preliminary offline learning is needed. First, we present an augmented form of NN that can be used to approximate functions with discontinuities, since friction has a jump discontinuity at the origin.

3.1 Neural network approximation of functions with discontinuities

The friction nonlinearity appearing in industrial actuators was discussed in Chapter 2 and is shown in Figure 3.1.1. It is discontinuous at the origin. One of the most important properties of NNs for control purposes is the universal approximation property. Unfortunately, this property is generally proven for continuous functions. In most real industrial control systems, there are nonsmooth functions (e.g., piecewise continuous) for which approximation results in the literature are sparse. Examples include friction, deadzones, and backlash. It is found that attempts to approximate piecewise continuous functions using smooth activation functions require many NN hidden nodes and many training iterations, and they still do not yield very good results.

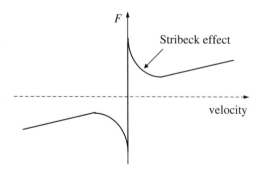

Figure 3.1.1. *Friction.*

Our main result in this section is an approximation theorem for piecewise continuous functions or functions with jumps. It is found that to approximate such functions suitably, it is necessary to augment the standard NN that uses smooth activation functions with *extra nodes* containing a certain *jump function approximation basis set* of nonsmooth activation functions. This novel augmented NN structure allows approximation of piecewise continuous functions of the sort that appear in motion control actuator nonlinearities. The augmented NN consists of neurons having standard sigmoid activation functions, plus some *additional neurons* having a special class of nonsmooth activation functions termed "jump function approximation basis functions." This augmented NN with additional neurons can approximate any piecewise continuous function with discontinuities at a finite number of known points.

3.1.1 Neural network approximation of continuous functions

Many well-known results say that any sufficiently smooth function can be approximated arbitrarily closely on a compact set using a two-layer NN with appropriate weights (Cybenko (1989), Hornik, Stinchcombe, and White (1989)). For instance, Cybenko's result for continuous function approximation says that given any function $f \in C(S)$ (the space of continuous functions), with S a compact subset of $\Re^n$, and any $\varepsilon > 0$, there exists a sum $G(x)$ of the form

$$G(x) = \sum_{k=0}^{L} \alpha_k \sigma(m_k^T x + n_k) \quad (3.1.1)$$

for some L, $m_k \in \Re^n$, $n_k \in \Re$, and $\alpha_k \in \Re$ such that

$$|G(x) - f(x)| < \varepsilon \quad (3.1.2)$$

for all $x \in S$. Generally, if ε is smaller, then the number L of hidden-layer neurons is larger. That is, a larger NN is required to approximate more accurately.

The function $\sigma(.)$ could be any continuous sigmoidal function, where a sigmoidal function is defined as

$$\sigma(x) \to \begin{cases} 1 & \text{for } x \to +\infty, \\ 0 & \text{for } x \to -\infty. \end{cases} \quad (3.1.3)$$

This result shows that any continuous function can be approximated arbitrarily well using a linear combination of sigmoidal functions. This is known as the NN *universal approximation property*.

3.1.2 Neural network approximation of functions with jumps

Industrial systems contain discontinuous nonlinearities, including friction and deadzones. There are a few results on approximation of nonsmooth functions (Park and Sandberg (1992)). Here are presented our main results for approximation of piecewise continuous functions or functions with jumps. It is found that to approximate such functions suitably, it is necessary to augment the set of functions $\sigma(.)$ used for approximation. In addition to continuous sigmoidal functions, one requires a set of *discontinuous basis functions*. We propose two suitable sets, the *polynomial jump approximation functions* and the *sigmoidal jump approximation functions*.

Theorem 3.1.1 (Approximation of piecewise continuous functions). *Let there be given any bounded function* $f : S \to \Re$ *that is continuous and analytic on a compact set* $S \subset \Re$ *except at* $x = c$, *where the function* f *has a finite jump and is continuous from the right. Then given any* $\varepsilon > 0$, *there exists a sum* F *of the form*

$$F(x) = g(x) + \sum_{k=0}^{N} a_k f_k(x - c) \tag{3.1.4}$$

such that

$$|F(x) - f(x)| < \varepsilon \tag{3.1.5}$$

for every x *in* S, *where* $g(.)$ *is a function in* $C^{\infty}(S)$ *and the* polynomial jump approximation basis functions f_k are defined as

$$f_k(x) = \begin{cases} 0 & for\ x < 0, \\ \frac{x^k}{k!} & for\ x \geq 0. \end{cases} \tag{3.1.6}$$

Proof. Let f be a smooth bounded function on S except at $x = c$, where the function has a finite discontinuity. Let g be an analytic function in $C^{\infty}(S)$ such that $f(x) = g(x)$ for all $x < c$, as shown in Figure 3.1.2.

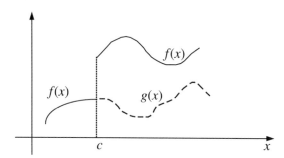

Figure 3.1.2. *Functions* f *and* g.

Then for $a \geq c$ and $x \geq c$, expand f and g into Taylor series as

$$f(x) = f(a) + \frac{f'(a)}{1!}(x - a) + \frac{f''(a)}{2!}(x - a)^2 + \cdots + \frac{f^{(N)}(a)}{n!}(x - a)^N + R_f^N(x, a),$$

$$g(x) = g(a) + \frac{g'(a)}{1!}(x - a) + \frac{g''(a)}{2!}(x - a)^2 + \cdots + \frac{g^{(N)}(a)}{n!}(x - a)^N + R_g^N(x, a).$$

Combining these two equations yields

$$f(x) = g(x) + f(a) - g(a) + \frac{f'(a) - g'(a)}{1!}(x - a) + \frac{f''(a) - g''(a)}{2!}(x - a)^2$$
$$+ \cdots + \frac{f^{(N)}(a) - g^{(N)}(a)}{N!}(x - a)^N + R_f^N(x, a) - R_g^N(x, a).$$

Letting $a \to c$ and knowing that the first N derivatives of g are continuous and that f is continuous from the right results in

$$f(x) = g(x) + f(c) - g(c) + \frac{f'(c) - g'(c)}{1!}(x - c) + \frac{f''(c) - g''(c)}{2!}(x - c)^2$$
$$+ \cdots + \frac{f^{(N)}(c) - g^{(N)}(c)}{N!}(x - c)^N + R_f^N(x, c) - R_g^N(x, c)$$

for $x \geq c$. Therefore, one has

$$f(x) = g(x) + a_0 f_0(x - c) + a_1 f_1(x - c) + \cdots + a_N f_N(x - c) + R_f^N(x, c) - R_g^N(x, c).$$

Since $R_f^N(x, c)$ and $R_g^N(x, c)$ go to zero as N approaches infinity, the proof is complete. □

We use the terminology "basis set" somewhat loosely. Note that $f_k(x - c)$ is a basis set for approximation of functions with discontinuities at a specified value $x = c$.

The next results follow directly from the foregoing proof.

Corollary 3.1.2. *Given the hypotheses of Theorem 3.1.1, define the* jump approximation error *as $E = |F(x) - f(x)|$. Then*

$$E = R_f^N(x, c) - R_g^N(x, c) = \frac{f^{(N+1)}(d)}{(N + 1)!}(x - c)^{N+1} - \frac{g^{(N+1)}(d)}{(N + 1)!}(x - c)^{N+1}, \quad (3.1.7)$$

where d is between c and x. Therefore, the error is bounded by

$$E \leq \frac{(x - c)^{(N+1)}}{(N + 1)!} \sup_d |f^{(N+1)}(d) - g^{(N+1)}(d)|, \quad (3.1.8)$$

where d is between c and x. □

The polynomial jump approximation basis functions f_k have a deficiency for feedback-control purposes in that they are not bounded. This could yield unbounded instabilities in a feedback control system. Therefore, it is now desired to replace these functions by another set of basis functions that are bounded. This yields a result more useful for closed-loop feedback-control purposes. To accomplish this, the following lemmas are needed.

Lemma 3.1.3 (Technical lemma).

$$\sum_{i=0}^{m} \binom{m}{i} (-1)^i (-i)^k = 0 \quad (3.1.9)$$

for every $m > k$, where $\binom{m}{i}$ is defined as usual,

$$\binom{m}{i} = \frac{m!}{i!(m - i)!}.$$

Proof. For the proof, see Selmic (2000). □

There follows the main result that allows us to replace the polynomial jump approximation basis set by discontinuous sigmoids.

Lemma 3.1.4 (Sigmoid jump approximation functions). *Any linear combination of polynomial jump approximation functions f_k, $\sum_{k=0}^{N} a_k f_k(x)$ can be represented as*

$$\sum_{k=0}^{N} a_k f_k(x) = z(x) + \sum_{k=0}^{N} b_k \varphi_k(x), \quad (3.1.10)$$

where the sigmoid jump approximation functions φ_k *are defined as*

$$\varphi_k(x) = \begin{cases} 0 & \text{for } x < 0, \\ (1 - e^{-x})^k & \text{for } x \geq 0, \end{cases} \tag{3.1.11}$$

and where $z(x)$ is a function that belongs to $C^N(S)$.

Proof. It is enough to prove that there exist coefficients b_k such that first N derivatives of the expression

$$\sum_{k=0}^{N} a_k f_k(x) - \sum_{k=0}^{N} b_k \varphi_k(x) \tag{3.1.12}$$

are continuous.

For $x < 0$, this expression has the constant value zero. Therefore, one must show that there exist coefficients b_i such that for $x > 0$, the first N derivatives of the expression are equal to zero.

For $x > 0$, one has

$$\sum_{k=0}^{N} a_k f_k(x) - \sum_{k=0}^{N} b_k \varphi_k(x) = \sum_{k=0}^{N} a_k \frac{x^k}{k!} - \sum_{k=0}^{N} b_k (1 - e^{-x})^k.$$

Using the binomial formula, one can expand this expression into

$$\sum_{k=0}^{N} a_k \frac{x^k}{k!} - \sum_{k=0}^{N} b_k \left[\sum_{i=0}^{k} \binom{k}{i} (-e^{-x})^i \right]$$

$$= \sum_{k=0}^{N} a_k \frac{x^k}{k!} - \sum_{k=0}^{N} b_k \left[\sum_{i=0}^{k} \binom{k}{i} (-1)^i (e^{-ix}) \right]$$

$$= \sum_{k=0}^{N} a_k \frac{x^k}{k!} - \sum_{k=0}^{N} b_k \left[\sum_{i=0}^{k} \binom{k}{i} (-1)^i \sum_{j=0}^{\infty} \frac{(-ix)^j}{j!} \right]$$

$$= \sum_{k=0}^{N} a_k \frac{x^k}{k!} - \sum_{k=0}^{N} \sum_{i=0}^{k} b_k \binom{k}{i} (-1)^i \sum_{j=0}^{\infty} (-i)^j \frac{x^j}{j!}$$

$$= \sum_{k=0}^{N} a_k \frac{x^k}{k!} - \sum_{k=0}^{N} \sum_{i=0}^{k} \sum_{j=0}^{\infty} b_k \binom{k}{i} (-1)^i (-i)^j \frac{x^j}{j!}$$

$$= \sum_{k=0}^{N} a_k \frac{x^k}{k!} - \sum_{j=0}^{\infty} \sum_{k=0}^{N} \sum_{i=0}^{k} b_k \binom{k}{i} (-1)^i (-i)^j \frac{x^j}{j!}$$

$$= \sum_{k=0}^{N} a_k \frac{x^k}{k!} - \sum_{k=0}^{\infty} \sum_{m=0}^{N} \sum_{i=0}^{m} b_m \binom{m}{i} (-1)^i (-i)^k \frac{x^k}{k!}.$$

In order for the first n derivatives of expression (3.1.12) to be zero, the coefficients of x^k for $k = 0, 1, \ldots, N$ should be zero in this equation. Therefore, one requires

$$a_k - \sum_{m=0}^{N} \sum_{i=0}^{m} b_m \binom{m}{i} (-1)^i (-i)^k = 0,$$

$$a_k - \sum_{m=0}^{N} b_m \sum_{i=0}^{m} \binom{m}{i} (-1)^i (-i)^k = 0.$$

Using Lemma 3.1.3, which says that

$$\sum_{i=0}^{m} \binom{m}{i}(-1)^i(-i)^k = 0$$

for every $m > k$, one obtains

$$a_k - \sum_{m=0}^{k} b_m \sum_{i=0}^{m} \binom{m}{i}(-1)^i(-i)^k = 0$$

for $k = 1, 2, \ldots, N$, or

$$a_k - b_k \sum_{i=0}^{k} \binom{k}{i}(-1)^i(-i)^k - \sum_{m=0}^{k-1} b_m \sum_{i=0}^{m} \binom{m}{i}(-1)^i(-i)^k = 0$$

for $k = 1, 2, \ldots, N$. Therefore, one obtains the recurrent relation for the coefficients b_k,

$$b_k = \frac{a_k - \sum_{m=0}^{k-1} b_m \sum_{i=0}^{m} \binom{m}{i}(-1)^i(-i)^k}{\sum_{i=0}^{k} \binom{k}{i}(-1)^i(-i)^k},$$

for $k = 1, 2, \ldots, N$ and with $b_0 = a_0$. $\Box$

The sigmoid jump approximation functions φ_k for $k = 0, 1, 2, 3$ are shown in Figure 3.1.3. These have a major advantage over the polynomial basis set for feedback controls design as they are bounded.

The next main result follows directly. It uses the sigmoid jump approximation basis set for approximation of jump functions by NNs.

Theorem 3.1.5 (General NN approximation result for functions with jumps). *Let there be given a bounded function $f : S \rightarrow \Re$, which is continuous and analytic on a compact set S except at $x = c$, where the function f has a finite jump and is continuous from the right. Given any $\varepsilon > 0$, there exists a sum F of the form*

$$F(x) = \sum_{k=0}^{L} \alpha_k \sigma(m_k x + n_k) + \sum_{k=0}^{N} a_k \varphi_k(x - c) \qquad (3.1.13)$$

such that

$$|F(x) - f(x)| < \varepsilon \qquad (3.1.14)$$

for every x in S, where $\sigma(x)$ is a standard sigmoid function and $\varphi_k(.)$ are the sigmoidal jump approximation functions. $\Box$

This theorem shows that any function with jumps can be approximated by a standard NN having standard continuous sigmoids (the first term of (3.1.13)) plus discontinuous sigmoidal jump functions (the second term of (3.1.13)).

It can be shown that, instead of the sigmoidal jump approximation functions, one can use either

$$\varphi_k(x) = \begin{cases} 0 & \text{for } x < 0, \\ \left(\dfrac{1 - e^{-x}}{1 + e^{-x}}\right)^k & \text{for } x \geq 0 \end{cases} \qquad (3.1.15)$$

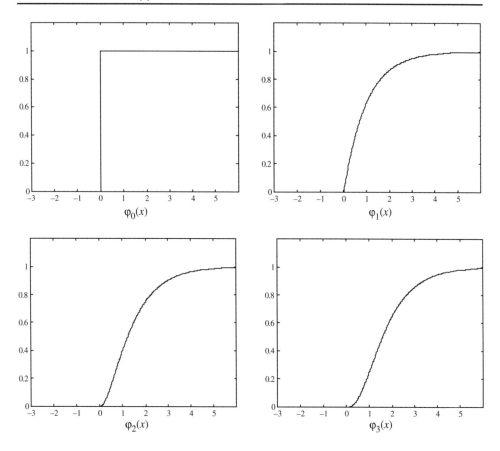

Figure 3.1.3. *Sigmoid jump approximation functions.*

or jump basis functions based on the hyperbolic tangent

$$\varphi_k(x) = \begin{cases} 0 & \text{for } x < 0, \\ \left(\dfrac{e^x - e^{-x}}{e^x + e^{-x}}\right)^k & \text{for } x \geq 0. \end{cases} \qquad (3.1.16)$$

3.1.3 Augmented multilayer neural networks for jump function approximation

Here we present the structure and training algorithms for the augmented multilayer NN just derived. This augmented NN is capable of approximating functions with jumps, provided that the points of discontinuity are known. Since the points of discontinuity are known in many nonlinear characteristics in industrial motion systems (e.g., the friction model is discontinuous at zero, the deadzone inverse is discontinuous at zero, etc.), this augmented NN is a useful tool for compensation of parasitic effects and actuator nonlinearities in industrial control systems.

The augmented multilayer NN is shown in Figure 3.1.4. One can see that the augmented NN is a combination of a standard NN (with standard activation functions $\sigma(.)$) plus some *extra neurons* with discontinuous sigmoidal activation functions φ_k. Comparing with the

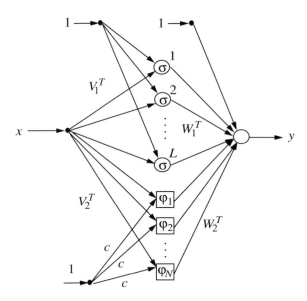

Figure 3.1.4. *Augmented multilayer NN.*

standard NN, one can see that the augmented NN has two sets of hidden activation functions, σ and φ_k, two sets of weights for the first layer, V_1^T and V_2^T, and two set of weights for the second layer, W_1^T and W_2^T. With this structure of NN, one has

$$y = \sum_{\ell=1}^{L} w_{1\ell}^1 \sigma(v_{\ell 1}^1 x_1 + v_{\ell 0}^1) + \sum_{\ell=1}^{N} w_{1\ell}^2 \varphi_\ell(v_{\ell 1}^2 x_1 + v_{\ell 0}^2) + w_{10}. \tag{3.1.17}$$

Similarly, as in the standard NN case, the output of the augmented NN is given in a matrix form as

$$y = W_1^T \sigma(V_1^T x) + W_2^T \varphi(V_2^T x). \tag{3.1.18}$$

If the hidden layer activation functions σ are defined as the sigmoids and one takes the jump basis functions φ_i as in (3.1.11), then using Theorem 3.1.5 says that this augmented NN can approximate any continuous function with a finite jump, provided that the jump location is known.

The standard NN tuning algorithms must be modified since the structure of the NN is changed. The set of second-layer weights W_1^T and W_2^T are trained as in the usual multilayer NN. Both of them are treated the same way, including threshold weights. The first-layer weights V_1^T are updated as in the usual NN (e.g., by a backpropagation approach). However, the weights V_2^T and the threshold c are fixed and are selected to correspond to the known point of discontinuity of the jump. Effectively, the jump approximation portion of the augmented NN is a one-layer NN, since it has only a single layer of tunable weights.

3.1.4 Simulation of the augmented multilayer neural network

Here are presented some simulation results that compare the approximation capabilities of the standard multilayer NN and our augmented NN in Figure 3.1.4. The activation functions

$\sigma(.)$ are taken as sigmoids, and the augmented NN has two additional jump approximation nodes with functions $\varphi_1(.)$, $\phi_2(.)$. The NNs were trained using backpropagation, with the weights V_2 in the augmented NN fixed according to the known point of discontinuity $c = 0$ of the function. Both NNs are trained for the same number of iterations. The NN toolbox in MATLAB was used (Demuth and Beale (1992)).

Neural network approximation of one-dimensional jump functions

Two examples of 1D jump functions are first chosen. In both cases, the NNs are selected to have $L = 20$ hidden nodes. The augmented NN has two additional nodes with sigmoidal jump approximation functions. Both NNs are first trained to approximate the discontinuous function defined as $y = \sin(x)$ for $x < -1$, else $y = 1 + \sin(x)$. The results are shown in Figures 3.1.5(a) and (b). Next, the function defined as $y = x$ for $x < 0$, else $y = 0.5x + 1$, was approximated. The results are shown in Figures 3.1.6(a) and (b). Note that in each case,

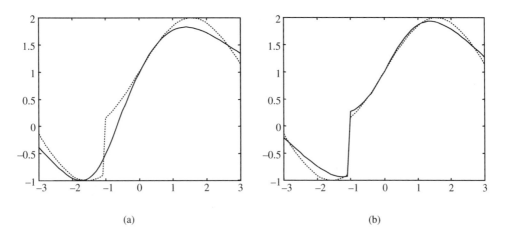

(a)   (b)

Figure 3.1.5. *Approximation of sinusoidal jump function using* (a) *standard NN and* (b) *augmented NN. NN function (full); desired function (dash).*

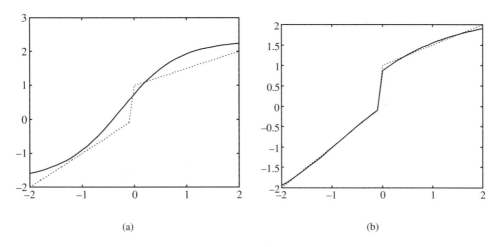

(a)   (b)

Figure 3.1.6. *Approximation of linear jump function using* (a) *standard NN and* (b) *augmented NN. NN function (full); desired function (dash).*

the augmented NN shown in part (b) does a better job of approximation since it can capture the jumps.

Neural network approximation of two-dimensional jump functions

Figures 3.1.7 and 3.1.8 show the superior performance of the augmented NN in approximating functions in two dimensions. The 2D function to be approximated is given by

$$f(x, y) = \begin{cases} y + 2 & \text{if } x > 0, \\ y - 2 & \text{if } x \leq 0. \end{cases}$$

This function is approximated using two NNs: the standard NN with continuous sigmoids and the augmented NN. Each NN has 25 hidden nodes, and the augmented NN has two additional nodes. Note that the error surface in the case of approximating with standard NNs is 10 times higher than the error surface when our augmented NN is used. This makes the augmented NN a very useful tool for friction and deadzone compensation, which will be used in our neurocontrol structure.

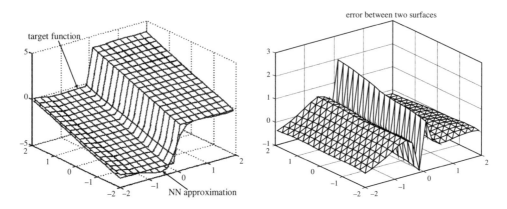

Figure 3.1.7. *Approximation of linear 2D jump function using standard NN.*

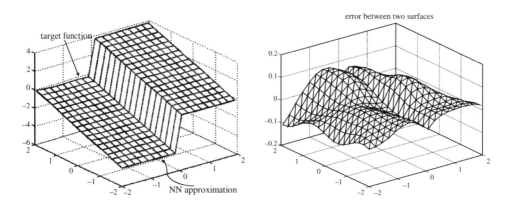

Figure 3.1.8. *Approximation of linear 2D jump function using augmented NN.*

The next simulation example shows the approximation capabilities of the augmented NN approximating the 2D function given by

$$f(x, y) = \begin{cases} y^2 + 2 & \text{if } x > 0, \\ y^2 - 2 & \text{if } x \leq 0. \end{cases}$$

For both NNs, $L = 20$, and the augmented NN has two additional nodes.

Figure 3.1.9 shows the result of approximation when the standard NN is used, and Figure 3.1.10 shows the result when the augmented NN is used. Once again the error surface is about 10 times smaller than the error surface when the standard NN is used.

From these simulations, it is obvious that the NN augmented by jump approximation functions has significantly better performance in approximating discontinuous functions and that it should be a very powerful tool for the compensation of actuator nonlinearities. In this simulation, the modified NN has only two jump nodes. Two or three jump nodes are usually enough in practical industrial applications.

3.2 Neurocontroller for friction compensation

Friction is a nonlinear effect that can limit the performance of industrial control systems; it occurs in all mechanical systems and therefore is unavoidable in control systems. It can cause

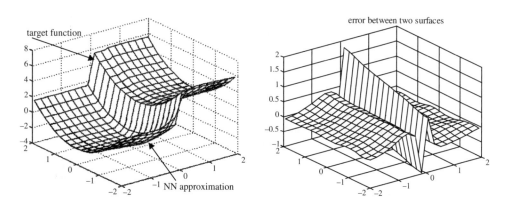

Figure 3.1.9. *Approximation of quadratic 2D jump function using standard NN.*

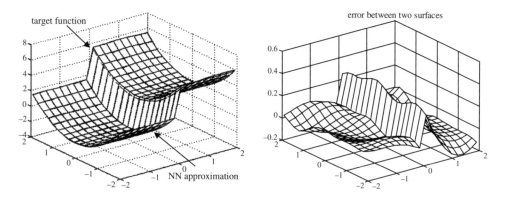

Figure 3.1.10. *Approximation of quadratic 2D jump function using augmented NN.*

tracking errors, limit cycles, and other undesirable effects. Friction is shown in Figure 3.1.1. Often, inexact friction compensation is used with standard adaptive techniques (Lee and Kim (1995)), which require models that are LIP. However, as shown in Chapter 2, realistic friction models are complex and are nonlinear in unknown parameters.

NNs have been used extensively in feedback control systems (Narendra and Partha-sarathy (1990)). Identification and compensation of the static friction model using NNs is performed in Du and Nair (1998) and Tzes, Peng, and Houng (1995) but without any closed-loop stability proof or guarantee of bounded weights. When stability proofs are provided, they rely almost invariably on the universal approximation property for NNs; see Chen and Khalil (1992), Lewis, Yesildirek, and Liu (1996), Polycarpou (1996), Rovithakis and Christodoulou (1994), Sadegh (1993), and Sanner and Slotine (1991). However, friction is a discontinuous function, and the augmented NN given in section 3.1 is far more suitable for use in industrial control systems for friction compensation than the standard NN based solely on continuous activation functions.

It is shown here how the augmented NN in section 3.1, capable of approximating piecewise continuous functions, can be used for friction compensation. A very general friction model is used. In contrast to standard adaptive control approaches, it is not required to use this approach for the model to be linear in the unknown parameters. We design a multiloop feedback neurocontroller for friction compensation that significantly improves the performance of industrial systems, and we provide a rigorous closed-loop system stability proof that guarantees small tracking error and bounded NN weights.

Mechanical industrial processes have a standard form known as the Lagrangian dynamics. We shall consider the robot arm dynamics as representative of such industrial motion processes.

3.2.1   Robot arm dynamics

In this section, we study the robot dynamics, deriving a suitable form for a controller that causes a robot manipulator to follow a desired trajectory. Then we design a neurocontroller that drives the robot along a prescribed trajectory while compensating for the deleterious effects of friction.

The dynamics of an n-link robot manipulator may be expressed in the Lagrange form (Lewis, Abdallah, and Dawson (1993))

$$M(\dot{q})\ddot{q} + V_m(q, \dot{q})\dot{q} + G(q) + F(\dot{q}) + \tau_d = \tau \qquad (3.2.1)$$

with $q(t) \in \Re^n$ the joint variable vector, $M(q)$ the inertia matrix, $V_m(q, \dot{q})$ the coriolis/centripetal matrix, $G(q)$ the gravity vector, and $F(\dot{q})$ the friction. Bounded unknown disturbances (including unstructured, unmodeled dynamics) are denoted by τ_d and the control input torque is $\tau(t)$.

Given a desired arm trajectory $q_d(t) \in \Re^n$, the tracking error is

$$e(t) = q_d(t) - q(t) \qquad (3.2.2)$$

and the filtered tracking error is

$$r = \dot{e} + \Lambda e, \qquad (3.2.3)$$

where $\Lambda = \Lambda^T > 0$ is a design parameter matrix. Differentiating $r(t)$, the arm dynamics may be written in terms of the filtered tracking error as

$$M\dot{r} = -V_m r - \tau + f + \tau_d, \qquad (3.2.4)$$

where the nonlinear robot function is

$$f(x) = M(\dot{q})(\ddot{q}_d + \Lambda\dot{e}) + V_m(q, \dot{q})(\dot{q}_d + \Lambda e) + G(q) + F(q, \dot{q}), \tag{3.2.5}$$

where

$$x \equiv \begin{bmatrix} q^T & \dot{q}^T & q_d^T & \dot{q}_d^T & \ddot{q}_d^T \end{bmatrix}^T. \tag{3.2.6}$$

The function $f(x)$ is generally unknown or incompletely known. It must be computed or compensated for by the feedback controller. Note that $f(x)$ is discontinuous since it contains unknown friction terms.

Define a control input torque as

$$\tau_0 = \hat{f} + K_v r \tag{3.2.7}$$

with gain matrix $K_v = K_v^T > 0$ and $\hat{f}(x)$ an estimate of $f(x)$. The structure of this control input is important since it determines the structure of the feedback control system. Note that it contains a proportional-plus-derivative (PD) control term

$$K_v r = K_v \dot{e} + K_v \Lambda e. \tag{3.2.8}$$

PD control is standard for the design of motion tracking loops for industrial systems. However, it also contains a compensation term $\hat{f}(x)$ that is going to be used to offset the deleterious effects of the unknown nonlinearities, including friction, by canceling them out. The control system must somehow provide this estimate $\hat{f}(x)$. In our case, an augmented NN will be used to learn the unknown system nonlinearities and provide this estimate.

The key to stability proofs in controller design is to examine the closed-loop error dynamics. The closed-loop error dynamics is given by

$$M\dot{r} = -(K_v + V_m)r + \tilde{f} + \tau_d, \tag{3.2.9}$$

where the functional estimation error is

$$\tilde{f} = f - \hat{f}. \tag{3.2.10}$$

Note that according to (3.2.3), one has

$$\frac{e}{r} = (sI + \Lambda)^{-1}, \qquad \frac{\dot{e}}{r} = s(sI + \Lambda)^{-1} \tag{3.2.11}$$

so that

$$\|e\| \le \|(sI + \Lambda)^{-1}\| \|r\|, \qquad \|\dot{e}\| \le \|s(sI + \Lambda)^{-1}\| \|r\|. \tag{3.2.12}$$

Using operator gain properties (Craig (1988)), it follows that

$$\|e\| \le \frac{\|r\|}{\sigma_{\min}(\Lambda)}, \qquad \|\dot{e}\| \le \|r\| \tag{3.2.13}$$

with $\sigma_{\min}(\Lambda)$ the minimum singular value of Λ (Lewis, Abdallah, and Dawson (1993)). This means that we can guarantee that both the tracking error $e(t)$ and its derivative $\dot{e}(t)$ are bounded simply by designing the control input (3.2.7) in such a way that the filtered error $r(t)$ is bounded. This is of crucial importance here because it reduces the complexity of the design problem. In fact, one notes that the original robot dynamics (3.2.1) is a dynamical system of order $2n$, while the filtered error system (3.2.9), which we can now use for design, is only of order n.

What remains in the controller design, to be tackled in the next subsection, is to select the PD gain K_v and, mainly, to find a technique for providing the estimate $\hat{f}(x)$ using an NN. There are several properties of the robot arm dynamics (Lewis, Abdallah, and Dawson (1993)) that are essential in these steps:

Property 1. $M(q)$ is a positive definite symmetric matrix bounded by $m_1 I \leq M(q) \leq m_2 I$, where m_1 and m_2 are known positive constants.

Property 2. $V_m(q, \dot{q})$ is bounded by $|V_m(q, \dot{q})| \leq v_b(q)\|\dot{q}\|$ with $v_b(q) : S \to \Re^1$ a known bound.

Property 3.   The matrix $\dot{M} - 2V_m$ is skew-symmetric.

Property 4.   The unknown disturbance satisfies $\|\tau_d\| < b_d$, where b_d is a known positive constant.

3.2.2   Neural network controller

Here is presented an NN robot controller based on a filtered-error approach and employing the augmented NN of section 3.1 to approximate the unknown nonlinear functions (3.2.25) in the robot arm dynamics, which include friction. A serious deficiency in using standard adaptive controllers in robotics is the requirement for *linearity in the unknown system parameters (LIP)*. LIP severely restricts the class of industrial systems that can be controlled, and indeed it does not hold for friction, as seen in Chapter 2. The LIP restriction is overcome here using the universal approximation property of NNs, which holds for general nonlinear functions. The main result presented here is going to be the controller in Figure 4.2.1 based on the results of Lewis, Yesildirek, and Liu (1996) but modified for friction compensation. Instead of requiring knowledge of the system structure, as in adaptive control, an NN is used to approximate the unmodeled dynamics. The use of the augmented NN gives the controller much better performance if there is significant unknown friction, since friction is discontinuous at zero.

In order to prove stability of the overall closed-loop system, some mild assumptions that hold in practical situations are required.

Assumption 1 (Bounded reference trajectory). The desired trajectory is bounded so that

$$\left\| \begin{matrix} q_d(t) \\ \dot{q}_d(t) \\ \ddot{q}_d(t) \end{matrix} \right\| \leq q_B \tag{3.2.14}$$

with q_B a known scalar bound.

Lemma 3.2.1 (Bound on NN input x). *For each time t, $x(t)$ is bounded by*

$$\|x\| \leq c_1 + c_2\|r\| \leq 2q_B + c_0\|r(0)\| + c_2\|r\| \tag{3.2.15}$$

for computable positive constants c_0, c_1, c_2.

Proof. The solution of LTI system (3.2.3) with the initial value vector $q(t_0)$ is

$$e(t) = e_0 \varepsilon^{-\Lambda(t-t_0)} + \int_{t_0}^{t} \varepsilon^{-\Lambda(t-\tau)} r(\tau) d\tau \quad \forall t \geq t_0,$$

where $e_0 = q_d(t_0) - q(t_0)$. Thus

$$\|e\| \leq \|e_0\| + \frac{\|r\|}{\sigma_{\min}(\Lambda)}$$

with $\sigma_{\min}(\Lambda)$ the minimum singular value of Λ. The NN input can be written as

$$x = \begin{bmatrix} q \\ \dot{q} \\ q_d \\ \dot{q}_d \\ \ddot{q}_d \end{bmatrix} = \begin{bmatrix} -e \\ -r + \Lambda e \\ q_d \\ \dot{q}_d \\ \ddot{q}_d \end{bmatrix} + \begin{bmatrix} q_d \\ \dot{q}_d \\ 0 \\ 0 \\ 0 \end{bmatrix}. \tag{3.2.16}$$

Then a bound can be given as

$$\|x\| \leq (1 + \sigma_{\max}(\Lambda))\|e\| + 2q_B + \|r\|$$
$$\leq \{[1 + \sigma_{\max}(\Lambda)]\|e_0\| + 2q_B\} + \left\{ 1 + \frac{1}{\sigma_{\min}(\Lambda)} + \frac{\sigma_{\max}(\Lambda)}{\sigma_{\min}(\Lambda)} \right\} \|r\|$$
$$= c_1 + c_2\|r\|$$

with

$$c_1 = [1 + \sigma_{\max}(\Lambda)]\|e_0\| + 2q_B,$$
$$c_2 = 1 + \frac{1}{\sigma_{\min}(\Lambda)} + \frac{\sigma_{\max}(\Lambda)}{\sigma_{\min}(\Lambda)}.$$

Now from (3.2.13), one has $\|e\| < \|r\|/\sigma_{\min}(\Lambda)$ for all t, whence one obtains that

$$c_0 = \frac{1 + \sigma_{\max}(\Lambda)}{\sigma_{\min}(\Lambda)}. \qquad \square$$

Neural network approximation and the nonlinearity in the parameters problem

The nonlinear robot function $f(x)$ given in (3.2.5) is a discontinuous function due to friction. Moreover, this function is unknown and must be estimated by the control system so it can be compensated for. Using the universal approximation property of NNs and our results for approximation of jump functions from section 3.1, there is a two-layer NN such that

$$f(x) = W_1^T \sigma(V_1^T x) + W_2^T \varphi(V_2^T x) + \varepsilon(x) \tag{3.2.17}$$

with the approximation error bounded on a compact set by $\|\varepsilon(x)\| < \varepsilon_N$, with ε_N a known bound.

The W_1, W_2, V_1, and V_2 are *ideal target weights* that give good approximations of $f(x)$. They are unknown and they may not even be unique. However, all that is required here is that they exist. In the derivation of the NN controller, then, it is up to us to show how to tune the NN weights so that they become close enough to the ideal weights.

Define $\hat{W}_1$, $\hat{W}_2$, $\hat{V}_1$, and $\hat{V}_2$ as estimates of the ideal NN weights, which are given by the NN tuning algorithms. Since the weights V_2^T are given by the designer depending on the location of the jumps, then $V_2^T = \hat{V}_2^T$. Define estimation error as

$$\tilde{W}_1 = W_1 - \hat{W}_1, \quad \tilde{W}_2 = W_2 - \hat{W}_2, \qquad \tilde{V}_1 = V_1 - \hat{V}_1. \tag{3.2.18}$$

The approximation of nonlinear function (3.2.5) is

$$\hat{f}(x) = \hat{W}_1^T \sigma(\hat{V}_1^T x) + \hat{W}_2^T \varphi(V_2^T x). \tag{3.2.19}$$

Let Z be the matrix of all the tuned NN weights

$$Z \equiv \begin{bmatrix} W_1 & 0 & 0 \\ 0 & W_2 & 0 \\ 0 & 0 & V_1 \end{bmatrix}. \tag{3.2.20}$$

Note that the weights V_2 are not included in (3.2.20) since they are known. They are selected according to the known location of the jump, which occurs at the origin for friction.

We require the next assumption, which always holds. It simply states that we can select a number large enough to overbound the unknown ideal NN weights.

Assumption 2 (Bounded ideal target NN weights). On any compact subset of $\Re^n$, the ideal NN weights are bounded so that

$$\|Z\|_F \le Z_B \tag{3.2.21}$$

with Z_B known.

We now show how to overcome the LIP problem, which has generally restricted the development of adaptive control tuning algorithms to systems that are LIP. The NN is nonlinear in the tunable parameters, since the tuned weights $\hat{V}_1$ are in the argument of the nonlinear activation functions.

Define the hidden-layer output error for a given x as

$$\tilde{\sigma} = \sigma(V_1^T x) - \sigma(\hat{V}_1^T x). \tag{3.2.22}$$

Using a Taylor series, one has

$$\sigma(V_1^T x) = \sigma(\hat{V}_1^T x) + \sigma'(\hat{V}_1^T x)\tilde{V}_1^T x + O(\tilde{V}_1^T x)^2, \tag{3.2.23}$$

where σ' denotes the first derivative and $O(\tilde{V}_1^T x)^2$ represents higher order terms. The importance of this equation is that it replaces $\tilde{\sigma}$, which is nonlinear in $\tilde{V}_1$, by an expression linear in $\tilde{V}_1$ plus higher order terms. The Jacobian $\sigma'(\hat{V}_1^T x)$ is easily computable in terms of measured signals $x(t)$ and the current weights $\hat{V}_1$.

It is shown by Lewis, Yesildirek, and Liu (1996) that the higher order terms in (3.2.23) are bounded by

$$\|O(\tilde{V}_1^T x)^2\| \le c_3 + c_4 q_B \|\tilde{V}_1\|_F + c_5 \|\tilde{V}_1\|_F \|r\| \tag{3.2.24}$$

for some constants c_3, c_4, c_5.

Controller structure and error system dynamics

The control input to the robot manipulator is selected as

$$\tau = \tau_0 - v = \hat{W}_1^T \sigma(\hat{V}_1^T x) + \hat{W}_2^T \varphi(V_2^T x) + K_v r - v, \qquad (3.2.25)$$

where the torque $\tau_0(t)$ is defined by (3.2.7). The estimate of the nonlinear function there has been provide by the NN as in (3.2.19). The function $v(t)$ is a robustifying term that is going to be selected later to help guarantee stability of the closed-loop system. The structure of this NN controller is shown in Figure 3.2.1. The neurocontroller is a multiloop feedback control system with an outer PD tracking loop and an inner NN loop that compensates for the unknown nonlinearities, including friction. Such an inner loop that cancels the unknown nonlinearities by estimating them is known as a feedback linearization loop.

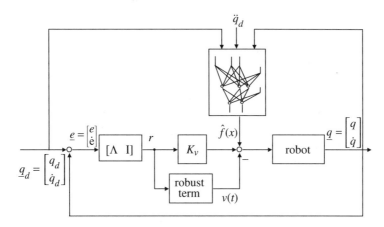

Figure 3.2.1. *NN controller.*

The closed-loop error dynamics is used for the stability proof. It is given by

$$\begin{aligned} M\dot{r} = {} & -(K_v + V_m)r + W_1^T \sigma(V_1^T x) + W_2^T \varphi(V_2^T x) \\ & - \hat{W}_1^T \sigma(\hat{V}_1^T x) - \hat{W}_2^T \varphi(V_2^T x) + \varepsilon + \tau_d + v. \end{aligned} \qquad (3.2.26)$$

Using (3.2.22), one has

$$M\dot{r} = -(K_v + V_m)r + \tilde{W}_1^T \sigma + W_1^T \tilde{\sigma} + \tilde{W}_2^T \varphi + \varepsilon + \tau_d + v. \qquad (3.2.27)$$

Applying the Taylor series approximation for $\tilde{\sigma}$, one can obtain the expression for closed-loop error dynamics (Lewis, Yesildirek, and Liu (1996)),

$$M\dot{r} = -(K_v + V_m)r + \tilde{W}_1^T (\hat{\sigma} - \hat{\sigma}' \hat{V}_1^T x) + \hat{W}_1^T \hat{\sigma}' \hat{V}_1^T x + \tilde{W}_2^T \varphi + w + v. \qquad (3.2.28)$$

The disturbance term comes from higher order terms in the Taylor series and is given by

$$w = \tilde{W}_1^T \hat{\sigma}' \hat{V}_1^T x + W_1^T O(\tilde{V}_1^T x)^2 + \varepsilon + \tau_d. \qquad (3.2.29)$$

It is shown by Lewis, Yesildirek, and Liu (1996) that this disturbance term is bounded by

$$\|w(t)\| \le C_0 + C_1 \|\tilde{Z}\|_F + C_2 \|\tilde{Z}\|_F \|r\| \qquad (3.2.30)$$

with C_i known positive constants.

Neural network weight tuning and stability analysis

The structure of the NN controller is given in Figure 3.2.1. We still need to show how to tune the NN weights in such a manner that closed-loop stability is guaranteed. Here we present NN weight-tuning laws, derived from Lyapunov theory, that guarantee stable tracking with internal stability. It is shown that the tracking error is suitably small, the NN weights are bounded, and the control $\tau(t)$ is bounded. The NN weights are adjusted online in real time with no preliminary offline learning required.

    The next assumption specifies the region of convergence of the two-layer NN controller to be derived.

Assumption 3 (Initial condition requirement). Suppose the desired trajectory $q_d, \dot{q}_d, \ddot{q}_d$ is bounded by q_B as in Assumption 1. Define the known constants c_0, c_2 by Lemma 4.2.2. Let the NN approximation property (3.2.17) hold for the unknown function $f(x)$ given in (3.2.5) with a given accuracy ε_N for all x in the compact set $S_x = \{x \mid \|x\| < b_x\}$ with $b_x > q_B$. Define $S_r = \{r \mid \|r\| < (b_x - q_B)/(c_0 + c_2)\}$. Let $r(0) \in S_r$.

    The set S_r specifies the set of allowed initial tracking errors $r(0)$. The intent is that the robot must initially be positioned near enough to the initial point of the desired trajectory. Note that the approximation accuracy of the NN determines the allowed magnitude of the initial tracking error $r(0)$. For a larger NN (i.e., more hidden-layer units), ε_N is small for a larger radius b_x; thus the allowed initial condition set S_r is larger. The key role of the initial condition requirement is in showing the dependence of the allowed initial condition set S_r on design parameters. The constants q_B, c_0, c_2, and b_x need not be explicitly determined.

    A key feature of the initial condition requirement is its independence of the NN initial weights. This is in stark contrast to other techniques in the literature where the proofs of stability depend on selecting some initial stabilizing NN weights, which is very difficult to do.

    The next main result shows how to select the robustifying term $v(t)$ and tune the NN in such a fashion that the closed-loop system is stable. It shows that the tracking error is UUB as defined in Chapter 2. This result completes the design of the friction compensation neurocontroller. It effectively shows that the standard backpropagation tuning algorithm must be modified for use in feedback control systems.

Theorem 3.2.2 (Augmented backprop weight tuning). *Let the desired trajectory be bounded by q_B as in Assumption 1 and the initial tracking error $r(0)$ satisfy initial condition Assumption 3. Let the ideal target weights be bounded as in Assumption 2. Take the control input for the robot dynamics as (3.2.25) with gain satisfying*

$$K_{V\,\min} > \frac{\left(C_0 + \frac{kC_3^2}{4}\right)(c_0 + c_2)}{b_x - q_B}, \tag{3.2.31}$$

where C_3 is defined in the proof and C_0 is defined in (3.2.30). Let the robustifying term be

$$v(t) = -K_z(\|\hat{Z}\|_F + Z_B r) \tag{3.2.32}$$

with gain

$$K_z > C_2. \tag{3.2.33}$$

Let the NN weight tuning be provided by

$$\dot{\hat{W}}_1 = S\hat{\sigma}r^T - S\hat{\sigma}'\hat{V}_1^T x r^T - kS\|r\|\hat{W}_1, \tag{3.2.34}$$

$$\dot{\hat{V}}_1 = Txr^T\hat{W}_1^T\hat{\sigma}' - kT\|r\|\hat{V}_1, \tag{3.2.35}$$

$$\dot{\hat{W}}_2 = E\varphi r^T - kE\|r\|\hat{W}_2 \tag{3.2.36}$$

with any constant matrices $S = S^T > 0$, $T = T^T > 0$, $E = E^T > 0$, and $k > 0$ a small scalar design parameter. Then the filtered tracking error $r(t)$ and NN weight estimates $\hat{V}_1$, $\hat{W}_1$, and $\hat{W}_2$ are UUB with bounds given by (3.2.37) and (3.2.38). Moreover, the tracking error may be kept as small as desired by increasing the gains K_v in (3.2.25).

Proof. Let us select a Lyapunov function as

$$L = \frac{1}{2}r^T M r + \frac{1}{2}\operatorname{tr}(\tilde{W}_1^T S^{-1}\tilde{W}_1) + \frac{1}{2}\operatorname{tr}(\tilde{V}_1^T T^{-1}\tilde{V}_1) + \frac{1}{2}\operatorname{tr}(\tilde{W}_2^T E^{-1}\tilde{W}_2).$$

One has the derivative

$$\dot{L} = r^T M\dot{r} + \frac{1}{2}r^T\dot{M}r + \operatorname{tr}(\tilde{W}_1^T S^{-1}\dot{\tilde{W}}_1) + \operatorname{tr}(\tilde{V}_1^T T^{-1}\dot{\tilde{V}}_1) + \operatorname{tr}(\tilde{W}_2^T E^{-1}\dot{\tilde{W}}_2).$$

Assuming that the initial tracking error satisfies initial condition Assumption 3, including the system error dynamics (3.2.28), one has

$$\dot{L} = -r^T(K_v + V_m)r + r^T\tilde{W}_1^T(\hat{\sigma} - \hat{\sigma}'\hat{V}_1^T x) + r^T\hat{W}_1^T\hat{\sigma}'\hat{V}_1^T x + r^T\tilde{W}_2^T\varphi + r^T w + r^T v$$
$$+ \frac{1}{2}r^T\dot{M}r + \operatorname{tr}(\tilde{W}_1^T S^{-1}\dot{\tilde{W}}_1) + \operatorname{tr}(\tilde{V}_1^T T^{-1}\dot{\tilde{V}}_1) + \operatorname{tr}(\tilde{W}_2^T E^{-1}\dot{\tilde{W}}_2).$$

Using Property 3 and the tuning rules yields

$$\dot{L} = -r^T K_v r + k\|r\|\operatorname{tr}\{\tilde{W}_1^T(W_1 - \tilde{W}_1)\} + k\|r\|\operatorname{tr}\{\tilde{V}_1^T(V_1 - \tilde{V}_1)\}$$
$$+ k\|r\|\operatorname{tr}\{\tilde{W}_2^T(W_2 - \tilde{W}_2)\} + r^T(w + v)$$
$$= -r^T K_v r + k\|r\|\operatorname{tr}\{\tilde{Z}^T(Z - \tilde{Z})\} + r^T(w + v).$$

Define

$$C_3 = Z_B + \frac{C_1}{k}.$$

Proceeding as in Lewis, Yesildirek, and Liu (1996), one obtains that $\dot{L}$ is negative as long as either

$$\|r\| > \frac{C_0 + \frac{kC_3^2}{4}}{K_{v_{\min}}} \equiv b_r \tag{3.2.37}$$

or

$$\|\tilde{Z}\|_F > \frac{C_3}{2} + \sqrt{\frac{C_0}{k} + \frac{C_3^2}{4}} \equiv b_Z. \tag{3.2.38}$$

Therefore, $\dot{L}$ is negative outside a compact set and $\|r\|$ and $\|\tilde{Z}\|_F$ are UUB as long as the control remains valid within this set. However, the PD gain limit (3.2.31) implies that the compact set defined by $\|r\| \leq b_r$ is contained in S_r so that the approximation property holds throughout.   □

In practice, it is not necessary to know the constants c_0, c_2, C_0, and C_3 to select the PD gain K_v according to (3.2.31). One must select only the gains large enough using some trial experiments or computer simulations. In practice, this has not shown itself to be a problem.

Some remarks will highlight the important features of our neurocontroller. A key feature of our neurocontroller is that no preliminary offline learning phase is required. Weight training occurs online in real time. Online training represents a greater advantage for the NN friction compensation scheme than offline training (Kovacic, Petik, and Bogdan (1999)) since the NN learns the friction during the natural controlled motions of the process.

Moreover, the NN weights are easy to initialize. The hidden-layer weights V_1 are initialized randomly, for it was shown in Igelnik and Pao (1995) that then the activation functions provide a basis. Note that the weights V_2 are not tuned since they are fixed and selected based on the known location of the discontinuity, which occurs at the origin for friction. The output-layer weights W_1 and W_2 are initialized at zero. In this way, the NN in Figure 3.2.1 is effectively absent at the beginning of training, so that the outer PD tracking loop stabilizes the system initially until the NN begins to learn.

The NN weight-tuning algorithms are an augmented form of backpropagation tuning and should be compared to the backprop algorithm given in Chapter 1. The first terms of (3.2.34), (3.2.35), and (3.2.36) are modified versions of the standard backpropagation algorithm where the *tracking error* is the backpropagated signal. The last terms correspond to the e-modification (e-mod) (Narendra and Annaswamy (1987)), which guarantees bounded NN weight estimates. It is important to note that one does not need the unknown plant Jacobian for these algorithms. All required quantities are easily computed in terms of signals easily measured in the feedback circuit.

The right-hand side of (3.2.37) can be taken as a practical bound on the tracking error in the sense that $r(t)$ will never stray far above it. It is important to note from this equation that the tracking error increases with the NN reconstruction error ε_N and robot disturbances b_d (both appear in C_0), yet arbitrarily small tracking errors may be achieved by selecting large gains K_V. On the other hand, (3.2.38) shows that the NN weight errors are fundamentally bounded by Z_B. In practical situations, we do not care whether the NN weights converge to the actual ideal weights that offer good approximation of the unknown robot function $f(x)$. The theorem shows that in a very practical sense, the tracking error $r(t)$ is nevertheless bounded to a small value. The tuning parameter k offers a design tradeoff between the relative eventual magnitudes of $\|r\|$ and $\|\tilde{Z}\|_F$.

Note that there is design freedom in the degree of complexity (e.g., size L) of the NN. For a more complex NN (e.g., more hidden units), the NN estimation error ε_N decreases, so the bounding constant C_0 will decrease, resulting in smaller tracking errors. On the other hand, a simplified NN with fewer hidden units will result in larger error bounds; this degradation can be compensated for by selecting a larger value for the PD gain K_v.

The proposed controller with friction compensation does not require LIP. Standard approaches to friction compensation usually require some previous linearization of the friction model in order to achieve LIP. The LIP requirement is a severe restriction for practical systems. Here the NN approximation property is used to avoid LIP. The nonlinear friction model is linear with respect to the nonlinear NN activation functions, which is a fundamental difference from the LIP condition.

3.2.3 Simulation of a neural network controller

To illustrate the performance of the augmented NN controller used for friction compensation, the two-link robot arm in Figure 3.2.2 is used. The dynamics of this arm are given by Lewis,

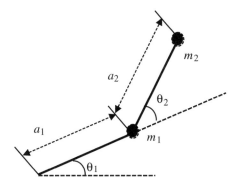

Figure 3.2.2. *Two-link robot arm.*

Abdallah, and Dawson (1993):

$$\tau_1 = [(m_1 + m_2)a_1^2 + m_2a_2^2 + 2m_2a_1a_2\cos\theta_2]\ddot{\theta}_1$$
$$+ [m_2a_2^2 + m_2a_1a_2\cos\theta_2]\ddot{\theta}_2 - m_2a_1a_2(2\dot{\theta}_1\dot{\theta}_2 + \dot{\theta}_2^2)\sin\theta_2$$
$$+ (m_1 + m_2)ga_1\cos\theta_1 + m_2ga_2\cos(\theta_1 + \theta_2) + \text{friction},$$
$$\tau_2 = [m_2a_2^2 + m_2a_1a_2\cos\theta_2]\ddot{\theta}_1 + m_2a_2^2\ddot{\theta}_2 + m_2a_1a_2\dot{\theta}_1^2\sin\theta_2$$
$$+ m_2ga_2\cos(\theta_1 + \theta_2) + \text{friction}.$$

The system parameters are chosen as $a_1 = 1.1, a_2 = 1, m_1 = 3$, and $m_2 = 2.3$. The nonlinear discontinuous static friction model from section 2.3 is included for both joints. The friction parameters for the first joint were taken as $\alpha_0 = 35$, $\alpha_1 = 1.1$, $\alpha_2 = 0.9$, $\beta_1 = 50$, and $\beta_2 = 65$, and those for the second joint were taken as $\alpha_0 = 38$, $\alpha_1 = 1$, $\alpha_2 = 0.95$, $\beta_1 = 55$, and $\beta_2 = 60$. It is assumed that the model of the robot arm as well as the friction model are unknown to the controller. Note that the friction is discontinuous at zero.

The NN weight-tuning parameters are chosen as $S = \text{diag}\{10, 10\}$, $T = \text{diag}\{10, 10\}$, $E = \text{diag}\{35, 35\}$, and $k = 0.01$. The NN controller parameters are chosen as $\Lambda = \text{diag}\{5, 5\}$ and $K_v = \{20, 20\}$. The input to the NN is given by (3.2.16). The NN has $L = 10$ hidden-layer units with sigmoidal activation functions and four additional nodes with jump approximation activation functions (two for each joint).

The selection of suitable NN parameters is not an easy issue. We performed the simulation several times with different values of these parameters and finally settled on the given values since they resulted in good responses. Specifically, the selection of the number of hidden-layer neurons is a thorny question. With $l = 8$, the response was degraded, and with $L = 12$ it was no better than using $L = 10$. Therefore, we settled on $L = 10$.

We compared three controllers. Controller 1 was a PD controller of the form of (3.2.25) but with the NN inner loop terms set to zero. This corresponds to using only the PD tracking loop in Figure 3.2.1. Controller 2 was the NN controller of Theorem 3.2.2 having an NN augmented with jump basis functions. Controller 3 was an NN controller like that in Theorem 3.2.2 but using a standard NN with no additional jump function approximation neurons.

Recall that the NN controller does not require preliminary offline training. We selected the initial weights $\hat{V}_1$ randomly, since it was shown in Igelnik and Pao (1995) that this provides a basis. Then the initial second-layer weights $\hat{W}_1$, $\hat{W}_2$ were set to zero. This means that at time $t = 0$, the NN is effectively absent in Figure 3.2.1 so that the PD loop keeps the system stable until the NN begins to learn and compensate for the friction effects.

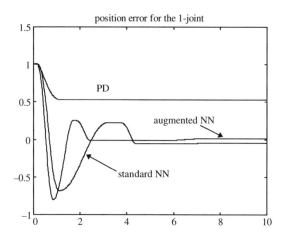

Figure 3.2.3. *Position error for 1-joint: PD controller, standard NN, and augmented NN.*

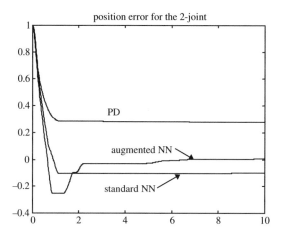

Figure 3.2.4. *Position error for 2-joint: PD controller, standard NN, and augmented NN.*

Figure 3.2.3 shows the position error for the first joint and Figure 3.2.4 for the second joint. A unit-step desired trajectory input is applied to the system. Notice that there is a steady-state error when PD control alone is used because of the presence of unknown dynamics as well as unknown friction. The standard NN controller reduces the steady-state error but it is still noticeable. This is due to the fact that the NN without augmented nodes for jump approximation cannot accurately approximate the friction, which has a jump at zero. The augmented NN controller shows far superior performance in compensating for friction.

The phase-plane plot in Figure 3.2.5 shows that the augmented NN does a superb job of bringing the steady-state error quickly to zero. Note that with PD control only, and also with the standard NN, the steady-state error is not equal to zero.

3.3   Conclusions

A new NN structure was presented for approximating piecewise continuous functions. A standard NN with continuous activation functions is augmented with an additional set of

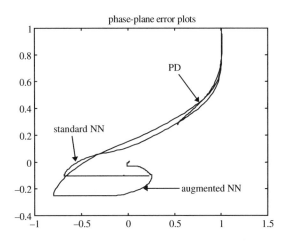

Figure 3.2.5. *Phase-plane error plots $e_2(t)$ versus $e_1(t)$.*

nodes with piecewise continuous activation functions. Friction is a discontinuous nonlinear function, and the augmented NN is used for its compensation. Based on the new augmented NN, it is not required for the friction model to be linear in unknown parameters, which is a major disadvantage in many existing adaptive control schemes.

An NN controller with friction compensation was designed based on the new augmented NN. The neurocontroller is a multiloop feedback control system with an outer tracking loop and an inner NN feedback linearization loop that compensates for the unknown nonlinearities, including friction. The NN controller does not require preliminary offline training but the weights are tuned online. The NN weights are easy to initialize. The first-layer weights are selected randomly, while the second-layer weights are set to zero values. This allows the outer PD tracking loop to stabilize the system until the NN begins to learn the nonlinearities and compensate for them.

A stability proof is given that does not require the selection of some initial stabilizing NN weights. An NN tuning law is provided that uses modified backpropagation tuning with additional *e*-mod terms.

Chapter 4

Neural and Fuzzy Control of Systems with Deadzones

In this chapter, we show how to use intelligent learning systems to compensate for the effects of deadzones in industrial motion systems. We discuss both NN and FL schemes for deadzone rejection. The NN and FL deadzone compensators must appear in the feedforward control loop, which presents difficulties in designing the controller structure and learning laws. The NN compensator uses two NNs, one to estimate the unknown deadzone and another to provide adaptive compensation in the feedforward path. The FL compensator is adaptive in the sense that its weights are tuned. Rigorous proofs of closed-loop stability for both deadzone compensators are provided and yield tuning algorithms for the weights of the NN or FL system. The first technique provides a general procedure for using NN to determine the preinverse of any unknown right-invertible function.

4.1 Introduction to deadzone control

Proportional-plus-derivative (PD) controllers are widely used in industrial and military systems. However, the motion actuators in such systems invariably have deadzone nonlinearities. Deadzone was discussed in Chapter 2. The deadzone nonlinearity is shown in Figure 4.2.1. Standard industrial control systems have been observed to result in limit cycles if the actuators have deadzones. The effects of deadzone are particularly deleterious in modern processes where precise motion and extreme speeds are needed, as in VLSI manufacturing, precision machining, and so on. Standard techniques for overcoming deadzone include variable structure control (Utkin (1978)) and dithering (Desoer and Shahruz (1986)). Rigorous results for motion tracking of such systems are notably sparse, though ad hoc techniques relying on simulations for verification of effectiveness are prolific.

Recently, in seminal work, several rigorously derived adaptive control schemes have been given for deadzone compensation (Tao and Kokotović (1996)). Compensation for nonsymmetric deadzones was considered for unknown linear systems in Tao and Kokotović (1994, 1995) and for nonlinear systems in Brunovsky form in Recker et al. (1991). Nonlinear Brunovsky form systems with unknown dynamics were treated in Tian and Tao (1996), where a backstepping approach was used. All of the known approaches to deadzone compensation using adaptive control techniques assume that the deadzone function can be linearly parametrized using a few parameters such as deadzone width, slope, and so forth. However, deadzones in industrial systems may not be linearly parametrizable.

Intelligent control techniques based on NN and FL systems have recently shown promise in effectively compensating for the effects of deadzone. NNs have been used extensively in feedback control systems. Most applications are ad hoc with no demonstrations of stability. The stability proofs that do exist rely almost invariably on the universal approximation property for NNs; see Chen and Khalil (1992), Lewis, Yesildirek, and Liu (1996), Polycarpou (1996), Sadegh (1993), and Sanner and Slotine (1991). However, to compensate for deadzone, one must estimate the deadzone inverse function, which is discontinuous at the origin. It is found that attempts to approximate jump functions using smooth activation functions require many NN nodes and many training iterations and still do not yield very good results.

In this chapter we show how to design a neurocontroller for deadzone compensation using the augmented NN from Chapter 3 to approximate the deadzone inverse. An NN scheme for deadzone compensation appears in Lee and Kim (1994), but no proof of performance is offered. We give rigorous stability proofs showing that the proposed controller affords closed-loop stability and much improved performance. The deadzone compensator consists of two NNs, one used as an estimator of the nonlinear deadzone function and the other used for the compensation itself. The NNs appear in the *feedforward* path, not in the feedback path as in other work cited above. In effect, the NNs learn the properties of the deadzone in real time so that the deadzone effect can be eliminated.

4.2   Position tracking controller with neural network deadzone compensation

In this section, we shall use NNs to design an NN compensator for industrial motion systems with deadzone nonlinearities. More details appear in Selmic and Lewis (2000).

4.2.1   Neural network deadzone precompensator

First, we are going to show how to use NNs to compensate for the deleterious effects of deadzone. We will end up with an NN precompensator, which will be used in subsequent sections to build a feedback neurocontroller for processes with deadzones.

Figure 4.2.1 shows a nonsymmetric deadzone nonlinearity $D(u)$, where u and τ are scalars. In general, u and τ are vectors. It is assumed that the deadzone has a nonlinear form.

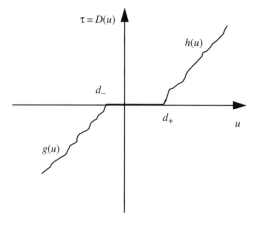

Figure 4.2.1. *Nonsymmetric deadzone nonlinearity.*

A mathematical model for the deadzone characteristic of Figure 4.2.1 is given by

$$\tau = D(u) = \begin{cases} g(u) < 0, & u \leq d_-, \\ 0, & d_- < u < d_+, \\ h(u) > 0, & u \geq d_+. \end{cases} \tag{4.2.1}$$

The functions $h(u)$ and $g(u)$ are smooth, nonlinear functions, so this describes a very general class of $D(u)$. All of $h(u)$, $g(u)$, d_+, and d_- are unknown, so compensation is difficult. We will require the functions $h(u)$ and $g(u)$ to be monotonically increasing, or the following.

Assumption 1. The functions $h(u)$ and $g(u)$ are smooth and invertible continuous functions.

In the case of symmetric deadzone, one has $d_+ = d_-$. If the functions $h(u)$ and $g(u)$ are linear, the deadzone is said to be linear. These restrictions may not apply in general industrial motion systems.

To offset the deleterious effects of deadzone, one may place a precompensator (as illustrated in Figure 4.2.4). There the desired function of the precompensator is to cause the composite throughput from w to τ to be unity. In order to accomplish this, it is necessary to generate the preinverse of the deadzone nonlinearity (Recker et al. (1991), Tao and Kokotović (1996)). By assumption, the function (4.2.1) is preinvertible. (Note that there is no postinverse for the deadzone.) Therefore, there exists a deadzone preinverse $D^{-1}(w)$ such that

$$D(D^{-1}(w)) = w. \tag{4.2.2}$$

The function $D^{-1}(w)$ is shown in Figure 4.2.2.

The mathematical model for the function shown in Figure 4.2.2 is given by

$$D^{-1}(w) = \begin{cases} g^{-1}(w), & w < 0, \\ 0, & w = 0, \\ h^{-1}(u), & w > 0. \end{cases} \tag{4.2.3}$$

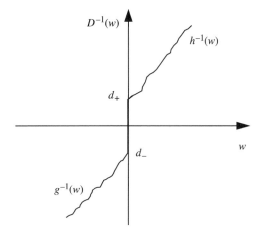

Figure 4.2.2. *Deadzone preinverse.*

The deadzone inverse $D^{-1}(w)$ can be expressed in equivalent form as

$$D^{-1}(w) = w + w_{\mathrm{NN}}(w), \tag{4.2.4}$$

where the *modified deadzone inverse* w_{NN} is given by

$$w_{\mathrm{NN}}(w) = \begin{cases} g^{-1}(w) - w, & w < 0, \\ 0, & w = 0, \\ h^{-1}(u) - w, & w > 0. \end{cases} \tag{4.2.5}$$

Equation (4.2.4) has a direct feedforward term plus a correction term. The function $w_{\mathrm{NN}}(w)$ is discontinuous at zero. The decomposition (4.2.4) of the deadzone preinverse is shown in Figure 4.2.3. This decomposition is important later in the design of the control system.

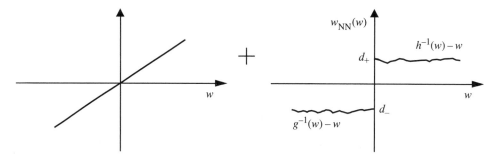

Figure 4.2.3. *Decomposition of deadzone inverse into direct feed term plus unknown discontinuous function.*

We are now going to introduce *two* NNs, one to approximate the deadzone and one its inverse. Based on the NN approximation property, one can approximate the deadzone function by

$$\tau = D(u) = W^T \sigma(V^T u + v_0) + \varepsilon(u). \tag{4.2.6}$$

Using the augmented NN with extra jump basis activation functions given in Chapter 3, one can design an NN for the approximation of the modified inverse function given in (4.2.5) by

$$w_{\mathrm{NN}}(w) = W_i^T \sigma(V_i^T w + v_{0i}) + \varepsilon_i(u). \tag{4.2.7}$$

In these equations $\varepsilon(u)$ and $\varepsilon_i(w)$ are the NN reconstruction errors and W and W_i are ideal target weights.

The reconstruction errors are bounded by $\|\varepsilon\| < \varepsilon_N(x)$ and $\|\varepsilon_i\| < \varepsilon_{Ni}(x)$, where x is equal to u and w, respectively. Here we consider x restricted to a compact set, and in that case these bounds are constant (i.e., $\varepsilon_N(x) = \varepsilon_N$, $\varepsilon_{Ni}(x) = \varepsilon_{Ni}$). The case where these bounds are not constant and not restricted to the compact set is treated in Lewis, Liu, and Yesildirek (1995), and in that case an additional saturation term has to be added to the controller. The first-layer weights V, V_i, v_0, and v_{0i} in both (4.2.6) and (4.2.7) are taken as fixed, and if they are properly chosen, the approximation property of the NN is still valid (Igelnik and Pao (1995)). In fact, the portion of the weights V_i corresponding to the jump approximation basis vectors (Chapter 3) are simply chosen based on the known location of the jump at the origin in the deadzone inverse.

We propose the structure of the NN deadzone precompensator shown in Figure 4.2.4, where there are two NNs. The input to the industrial process is given by the signal $\tau(t)$, which we assume cannot be measured directly. We shall soon demonstrate the effectiveness of this neurocompensator for the deadzone. The first NN (NN I; see, e.g., (4.2.6)) is used as a deadzone estimator or observer, while the second (NN II) is used as a deadzone compensator. Note that only the output of NN II is directly affecting the input $u(t)$, while NN I is a higher level "performance evaluator" that is used for tuning NN II. As will be shown mathematically later, the precompensator actually inverts the nonlinear deadzone function. In fact, this dual NN compensator represents a technique for adaptively inverting any nonlinear right-invertible functions in industrial motion device actuators.

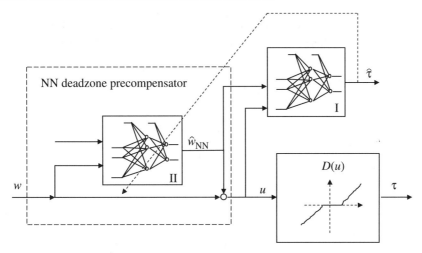

Figure 4.2.4. *Dual NN deadzone compensation scheme.*

We are now going to show that the dual NN precompensator effectively inverts the deadzone under certain conditions. To do this, it is necessary to focus on the transmission characteristics from $w(t)$ to $\tau(t)$, which we would like to be equal to unity.

Define $\hat{W}$ and $\hat{W}_i$ as estimates of the ideal NN weights. These are the actual values of weights appearing in the circuit and are given by the NN tuning algorithms. Define the weight estimation errors as

$$\tilde{W} = W - \hat{W}, \qquad \tilde{W}_i = W_i - \hat{W}_i \tag{4.2.8}$$

and the approximations of the nonlinear deadzone and modified deadzone inverse functions as

$$\hat{\tau} = \hat{D}(u) = \hat{W}^T \sigma(V^T u + v_0), \tag{4.2.9}$$

$$\hat{w}_{\text{NN}}(w) = \hat{W}_i^T \sigma(V_i^T w + v_{0i}). \tag{4.2.10}$$

Note that expressions (4.2.9) and (4.2.10) represent, respectively, an NN approximation of the deadzone function (4.2.1) and of the modified deadzone inverse (4.2.5). These are denoted, respectively, as NN I and NN II in Figure 4.2.4. The signal $\hat{w}_{\text{NN}}(w)$ is used for the deadzone compensation, and $\hat{\tau}$ represents the estimated value of signal τ, which we assume cannot be measured directly. We say that NN I is an *observer* for the unmeasurable process control input $\tau(t)$. Note that

$$u = w + \hat{w}_{\text{NN}}(w). \tag{4.2.11}$$

The next bounding assumption is needed and is always true.

Assumption 2 (Bounded ideal NN weights). The ideal weights W and W_i are bounded such that $\|W\|_F \leq W_M$ and $\|W_i\|_F \leq W_{iM}$, with W_M and W_{iM} known bounds.

The next result shows the effectiveness of the proposed NN structure, by providing an expression for the *composite throughput error* of the compensator plus deadzone. It shows that, as the estimates $\hat{W}$ and $\hat{W}_i$ approach the actual NN parameters W and W_i, the NN precompensator effectively provides a *preinverse* for the deadzone nonlinearity. Later it is shown how to tune (4.2.9) and (4.2.10) so that $\hat{W}$ and $\hat{W}_i$ are close to W and W_i.

Theorem 4.2.1 (Throughput error using dual NN deadzone compensation). *Given the NN deadzone compensator* (4.2.10), (4.2.11) *and the NN observer* (4.2.9), *the throughput of the compensator plus the deadzone is given by*

$$\tau = w - \hat{W}^T \sigma'(V^T u + v_0) V^T \tilde{W}_i^T \sigma_i(V_i^T w + v_{0i}) + \tilde{W}^T \sigma'(V^T u + v_0) V^T \hat{w}_{\mathrm{NN}} + d(t),$$
$$(4.2.12)$$

where the modeling mismatch term $d(t)$ is given by

$$d(t) = -\tilde{W}^T \sigma'(V^T u + v_0) V^T W_i^T \sigma_i(V_i^T w + v_{0i}) - b(t) + \varepsilon(u) \qquad (4.2.13)$$

with $b(t)$ defined in (4.2.16).

Proof. From (4.2.6) and (4.2.11), one has

$$\tau = W^T \sigma(V^T(w + \hat{w}_{\mathrm{NN}}) + v_0) + \varepsilon(w + \hat{w}_{\mathrm{NN}}). \qquad (4.2.14)$$

From (4.2.2), it follows that

$$w = W^T \sigma(V^T(w + w_{\mathrm{NN}}) + v_0) + \varepsilon(w + w_{\mathrm{NN}}),$$

whence by (4.2.7),

$$w = W^T \sigma[V^T(w + W_i^T \sigma_i(V_i^T w + v_{0i}) + \varepsilon_i(w)) + v_0] + \varepsilon(w + w_{\mathrm{NN}}),$$
$$w = W^T \sigma[V^T(w + \hat{W}_i^T \sigma_i(V_i^T w + v_{0i}) + \tilde{W}_i^T \sigma_i(V_i^T w + v_{0i}) + \varepsilon_i(w)) + v_0]$$
$$+ \varepsilon(w + w_{\mathrm{NN}}).$$

Using the Taylor series expansion, one has

$$w = W^T \sigma[V^T(w + \hat{W}_i^T \sigma_i(V_i^T w + v_{0i})) + v_0]$$
$$+ W^T \sigma'[V^T(w + \hat{W}_i^T \sigma_i(V_i^T w + v_{0i})) + v_0]V^T(\tilde{W}_i^T \sigma_i(V_i^T w + v_{0i}) + \varepsilon_i(w))$$
$$+ W^T R_1(\tilde{W}_i, w) + \varepsilon(w + w_{\mathrm{NN}}),$$

where $R_1(\tilde{W}_i, w)$ is the remainder of the first Taylor polynomial. Regrouping the terms, one has

$$w = W^T \sigma[V^T(w + \hat{W}_i^T \sigma_i(V_i^T w + v_{0i})) + v_0]$$
$$+ W^T \sigma'[V^T(w + \hat{W}_i^T \sigma_i(V_i^T w + v_{0i})) + v_0]V^T(\tilde{W}_i^T \sigma_i(V_i^T w + v_{0i})) + b(t),$$
$$(4.2.15)$$

where $b(t)$ is given by

$$b(t) = W^T \sigma'[V^T(w + \hat{W}_i^T \sigma_i(V_i^T w + v_{0i})) + v_0]V^T \varepsilon_i(w) + W^T R_1(\tilde{W}_i, w)$$
$$+ \varepsilon(w + w_{NN}).$$
(4.2.16)

Combining (4.2.15) and (4.2.11) gives

$$\begin{aligned} w + \varepsilon(u) &= W^T \sigma(V^T u + v_0) + \varepsilon(u) + \hat{W}^T \sigma'(V^T u + v_0)V^T(\tilde{W}_i^T \sigma_i(V_i^T w + v_{0i})) \\ &\quad + \tilde{W}^T \sigma'(V^T u + v_0)V^T(\tilde{W}_i^T \sigma_i(V_i^T w + v_{0i})) + b(t), \\ w + \varepsilon(u) &= W^T \sigma(V^T u + v_0) + \varepsilon(u) + \hat{W}^T \sigma'(V^T u + v_0)V^T(\tilde{W}_i^T \sigma_i(V_i^T w + v_{0i})) \\ &\quad + \tilde{W}^T \sigma'(V^T u + v_0)V^T(W_i^T \sigma_i(V_i^T w + v_{0i})) \\ &\quad - \tilde{W}^T \sigma'(V^T u + v_0)V^T(\hat{W}_i^T \sigma_i(V_i^T w + v_{0i})) + b(t), \end{aligned}$$

which combined with (4.2.14) gives (4.2.12).   □

The form of (4.2.12) will be crucial in deriving the NN tuning laws that guarantee closed-loop stability. The first term has known factors multiplying $\tilde{W}_i$, the second term has known factors multiplying $\tilde{W}$, and a suitable bound can be found for $d(t)$. The form of (4.2.12) is similar to the form in Tao and Kokotović (1994), but instead of parametrizing the direct and the inverse functions with the same parameters, we use different function approximators for the direct and inverse functions together with crucial fact (4.2.2), which is actually a connection between them. The expression (4.2.2) couples the information inherent in NN I and NN II. Intuitively, compensating the unknown effect (NN II) depends on what one observes (NN I), and vice versa: observation of the unknown effect (NN I) depends on how one modifies the system (NN II). This will later be very clearly seen in the tuning laws derived for $\tilde{W}_i$ and $\tilde{W}$, where the differential equations for tuning NN I and NN II are mutually coupled.

In general, this proposed NN compensation scheme could be used for inverting *any* continuous invertible function. Therefore, it is a powerful result for compensation of general actuator nonlinearities in motion control systems.

The next result gives us the upper bound of the norm of $d(t)$. It is an important result used in the stability proof for the control system in the next section.

Lemma 4.2.2 (Bound on the modeling mismatch term). *The norm of the modeling mismatching term $d(t)$ in (4.2.13) is bounded on a compact set by*

$$\|d(t)\| \le a_1 \|\tilde{W}\|_F + a_2 \|\tilde{W}_i\|_F^2 + a_3 \|\tilde{W}_i\|_F + a_5,$$
(4.2.17)

where a_1, a_2, a_3, a_4, a_5 are computable constants.

Proof. From (4.2.13), one has

$$\begin{aligned} \|d(t)\| &\le \|\tilde{W}\|_F \|\sigma'(\cdot)\| \|V\|_F \|W_i\|_F \|\sigma_i(\cdot)\| + \|b(t)\| + \|\varepsilon(w + \hat{w}_{NN})\|, \\ \|d(t)\| &\le \|\tilde{W}\|_F \|\sigma'(\cdot)\| \|V\|_F W_{iM} \|\sigma_i(\cdot)\| + \|b(t)\| + \varepsilon_N, \\ \|d(t)\| &\le a_1 \|\tilde{W}\|_F + \|b(t)\| + \varepsilon_N. \end{aligned}$$
(4.2.18)

From the definition (4.2.16), it follows that

$$
\begin{aligned}
\|b(t)\| &\leq \|W\|_F \|\sigma'(\cdot)\| \|V\|_F \|\varepsilon_i(w)\| + \|W\|_F \frac{1}{2} \|\sigma''(\cdot)\| \|V\|_F^2 \|\tilde{W}_i\|_F^2 \|\sigma_i(\cdot)\|^2 \\
&\quad + \|W\|_F \|\sigma''(\cdot)\| \|V\|_F^2 \|\tilde{W}_i\|_F \|\sigma_i(\cdot)\| \|\varepsilon_i(w)\| \\
&\quad + \|W\|_F \frac{1}{2} \|\sigma''(\cdot)\| \|V\|_F^2 \|\varepsilon_i(w)\|^2 + \|\varepsilon(w + w_{\mathrm{NN}})\|,
\end{aligned}
$$

$$
\begin{aligned}
\|b(t)\| &\leq W_M \|\sigma'(\cdot)\| \|V\|_F \varepsilon_{Ni} + W_M \frac{1}{2} \|\sigma''(\cdot)\| \|V\|_F^2 \|\tilde{W}_i\|_F^2 \|\sigma_i(\cdot)\|^2 \\
&\quad + W_M \|\sigma''(\cdot)\| \|V\|_F^2 \|\tilde{W}_i\|_F \|\sigma_i(\cdot)\| \varepsilon_{Ni} + W_M \frac{1}{2} \|\sigma''(\cdot)\| \|V\|_F^2 \varepsilon_{Ni}^2 + \varepsilon_N,
\end{aligned}
\tag{4.2.19}
$$

$$
\|b(t)\| \leq a_2 \|\tilde{W}_i\|_F^2 + a_3 \|\tilde{W}_i\|_F + a_4,
$$

where a_2, a_3, and a_4 are computable constants. Combining (4.2.18) and (4.2.19), one gets (4.2.17).   □

4.2.2   Robot arm dynamics

Mechanical industrial processes have a standard form known as the Lagrangian dynamics. We shall consider the robot arm dynamics as representative of such industrial motion processes. VLSI positioning tables, CNC (computer numerically controlled) machine tools, and so on have the same form of dynamics. As seen in Chapter 3, the dynamics of an n-link robot manipulator may be expressed in the Lagrange form (Lewis, Abdallah, and Dawson (1993))

$$
M(\dot{q})\ddot{q} + V_m(q, \dot{q})\dot{q} + G(q) + F(\dot{q}) + \tau_d = \tau
\tag{4.2.20}
$$

with $q(t) \in \Re^n$ being the joint variable vector, $M(q)$ the inertia matrix, $V_m(q, \dot{q})$ the coriolis/centripetal matrix, $G(q)$ the gravity vector, and $F(\dot{q})$ the friction. Bounded unknown disturbances (including unstructured, unmodeled dynamics) are denoted by τ_d, and the control input torque is $\tau(t)$.

Given a desired arm trajectory $q_d(t) \in \Re^n$, the tracking error is

$$
e(t) = q_d(t) - q(t)
\tag{4.2.21}
$$

and the filtered tracking error is

$$
r = \dot{e} + \Lambda e,
\tag{4.2.22}
$$

where $\Lambda = \Lambda^T > 0$ is a design parameter matrix. It was seen in Chapter 3 that if we can design a controller to keep $r(t)$ bounded, then $e(t)$ and its derivatives are also bounded. Therefore, we shall focus on keeping $r(t)$ small. This reduces the control problem from dimension $2n$ to dimension n, therefore affording a simpler design procedure.

Differentiating $r(t)$, the arm dynamics may be written in terms of the filtered tracking error as

$$
M\dot{r} = -V_m r - \tau + f + \tau_d,
\tag{4.2.23}
$$

where the nonlinear robot function is

$$
f(x) = M(\dot{q})(\ddot{q}_d + \Lambda \dot{e}) + V_m(q, \dot{q})(\dot{q}_d + \Lambda e) + G(q) + F(q, \dot{q}),
\tag{4.2.24}
$$

where

$$x \equiv [q^T \quad \dot{q}^T \quad q_d^T \quad \dot{q}_d^T \quad \ddot{q}_d^T]^T. \tag{4.2.25}$$

The function $f(x)$ is generally unknown or incompletely known. It must be computed or compensated for by the feedback controller. Note that this choice of the vector $x(t)$ is not unique since, for instance, one could replace $q(t)$ and $q_d(t)$ by $q(t)$ and the error $e(t)$.

There are several properties of the robot arm dynamics (Lewis, Abdallah, and Dawson (1993)) that are essential in the control design procedure of the next subsection:

Property 1. $M(q)$ is a positive definite symmetric matrix bounded by $m_1 I \leq M(q) \leq m_2 I$, where m_1 and m_2 are known positive constants.

Property 2. $V_m(q, \dot{q})$ is bounded by $|V_m(q, \dot{q})| \leq v_b(q)\|\dot{q}\|$ with $v_b(q) : S \to \Re^1$ a known bound.

Property 3.   The matrix $\dot{M} - 2V_m$ is skew-symmetric.

Property 4.   The unknown disturbance satisfies $\|\tau_d\| < \tau_M$, where τ_M is a known positive constant.

4.2.3   Design of tracking controller with neural network deadzone compensation

In this section, it is shown how to use the dual NN deadzone compensator just designed in a closed-loop feedback control system to guarantee motion tracking with deadzone compensation in robotic systems. Of course, we must prove closed-loop stability and also show how to tune the NN weights so that the actual weights are near the ideal weights, as required in Theorem 4.2.1 for the throughput error to be small.

A robust compensation scheme for unknown terms in the nonlinear robot function $f(x)$ is provided by selecting the tracking controller

$$w = \hat{f}(x) + K_V r - v \tag{4.2.26}$$

with $\hat{f}(x)$ an estimate for the nonlinear terms $f(x)$ and $v(t)$ a robustifying term to be selected for the disturbance rejection. The feedback gain matrix $K_V > 0$ is often selected diagonal.

There are several ways to select the estimate $\hat{f}(x)$. If $f(x)$ in (4.2.24) is unknown, it can be estimated using adaptive control techniques (Craig (1988), Lewis, Abdallah, and Dawson (1993), Slotine and Li (1988)) or the NN controller in Lewis, Yesildirek, and Liu (1996). In this section, we do not want to distract from the technique of NN deadzone compensation being presented, so we shall assume that the estimate $\hat{f}(x)$ is fixed and will not be adapted. That is, we are taking a robust control approach (Corless and Leitmann (1982), Lewis, Abdallah, and Dawson (1993)). We make the next assumption on the goodness of the estimate $\hat{f}(x)$ for $f(x)$.

Assumption 3 (Bounded estimation error). The nonlinear function $f(x)$ is assumed to be unknown, but a fixed estimate $\hat{f}(x)$ is assumed known such that the functional estimation error $\tilde{f}(x) = f(x) - \hat{f}(x)$ satisfies

$$\|\tilde{f}\| \leq f_M(x) \tag{4.2.27}$$

for some known bounding function $f_M(x)$.

This is a standard assumption in robust control (Corless and Leitmann (1982)) since in practical systems the bound $f_M(x)$ can be computed knowing the upper bound on payload masses, frictional effects, and so on.

Deadzone compensation is provided using

$$u = w + \hat{w}_{\mathrm{NN}} = w + \hat{W}_i^T \sigma_i (V_i^T w + v_{0i}), \tag{4.2.28}$$

where $\hat{W}_i$ and V_i are the weights of NN II in Figure 4.2.5. The multiloop control structure implied by this scheme is shown in Figure 4.2.5, where $\underline{q} = [q^T \quad \dot{q}^T]^T$, $\underline{q}_d = [q_d^T \quad \dot{q}_d^T]^T$, and $\underline{e} = [e^T \quad \dot{e}^T]^T$. The controller has a PD tracking loop with gains $K_V r = K_V \dot{e} + K_V \Lambda e$, and the deadzone effect is ameliorated by the NN feedforward compensator. The estimate $\hat{f}(x)$ is computed by an inner nonlinear control loop. This is known as a feedback linearization loop.

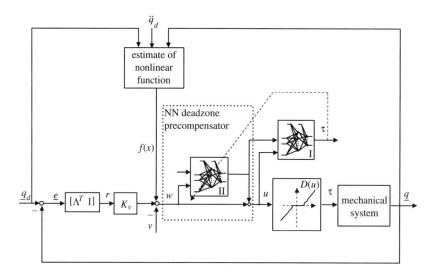

Figure 4.2.5. *Tracking controller with NN deadzone compensation.*

In order to design an NN system such that the tracking error $r(t)$ is bounded and all internal states are stable, one must examine the error dynamics. Substituting (4.2.26) and (4.2.12) into (4.2.23) yields the closed-loop error dynamics

$$\begin{aligned}
M\dot{r} = &-V_m r - K_V r + \hat{W}^T \sigma'(V^T u + v_0) V^T \tilde{W}_i^T \sigma_i (V_i^T w + v_{0i}) \\
&- \tilde{W}^T \sigma'(V^T u + v_0) V^T \hat{w}_{\mathrm{NN}} - d(t) + \tilde{f} + \tau_d + v,
\end{aligned} \tag{4.2.29}$$

where the nonlinear function estimation error is given by $\tilde{f}(x) = f(x) - \hat{f}(x)$.

The next assumption on the reference trajectory always holds in practice.

Assumption 4 (Bounded reference trajectory). The desired trajectory is bounded so that

$$\left\| \begin{matrix} q_d(t) \\ \dot{q}_d(t) \\ \ddot{q}_d(t) \end{matrix} \right\| \le q_B, \tag{4.2.30}$$

with q_B a known scalar bound.

Weight tuning and stability analysis

We have specified the NN controller shown in Figure 4.2.5. We are now ready to complete the design of the deadzone compensator by proving stability. This requires that the NN weights be tuned in a special way. The next theorem provides algorithms for tuning the NN weights for the deadzone precompensator with guaranteed closed-loop stability. The stability notion used is the practical definition of uniform ultimate boundedness presented in Chapter 2. Recall that $\hat{W}$ and V are the weights of NN I in Figure 4.2.5.

Theorem 4.2.3 (Tuning of NN adaptive deadzone compensator). *Given the system in (4.2.20) and Assumptions 1–4, select the tracking control law (4.2.26) plus the deadzone compensator (4.2.28). Choose the robustifying signal as*

$$v(t) = -(f_M(x) + \tau_M)\frac{r}{\|r\|}, \tag{4.2.31}$$

where the $f_M(x)$ and τ_M are bounds on the functional estimation error and disturbance, respectively. Let the estimated NN weights be provided by the NN tuning algorithm

$$\dot{\hat{W}} = -S\sigma'(V^T u + v_0)V^T \hat{W}_i^T \sigma_i(V_i^T w + v_{0i})r^T - k_1 S\|r\|\hat{W}, \tag{4.2.32}$$

$$\dot{\hat{W}}_i = T\sigma_i(V_i^T w + v_{0i})r^T \hat{W}^T \sigma'(V^T u + v_0)V^T - k_1 T\|r\|\hat{W}_i - k_2 T\|r\|\|\hat{W}_i\|_F \hat{W}_i, \tag{4.2.33}$$

with any constant matrices $S = S^T > 0$, $T = T^T > 0$, and $k_1, k_2 > 0$ small scalar design parameters. Then the filtered tracking error $r(t)$ and NN weight estimates $\hat{W}$ and $\hat{W}_i$ are UUB, with bounds given by (4.2.34), (4.2.35), and (4.2.36). Moreover, the tracking error may be kept as small as desired by increasing the gains K_v.

Proof. Select the Lyapunov function candidate

$$L = \frac{1}{2}r^T M r + \frac{1}{2}\operatorname{tr}[\tilde{W}^T S^{-1}\tilde{W}] + \frac{1}{2}\operatorname{tr}[\tilde{W}_i^T T^{-1}\tilde{W}_i].$$

Differentiating L yields

$$\dot{L} = r^T M\dot{r} + \frac{1}{2}r^T \dot{M}r + \operatorname{tr}[\tilde{W}^T S^{-1}\dot{\tilde{W}}] + \operatorname{tr}[\tilde{W}_i^T T^{-1}\dot{\tilde{W}}_i],$$

and using (4.2.29) yields

$$\dot{L} = -r^T V_m r - r^T K_V r + r^T \hat{W}^T \sigma'(V^T u + v_0)V^T \tilde{W}_i^T \sigma_i(V_i^T w + v_{0i})$$
$$- r^T \tilde{W}^T \sigma'(V^T u + v_0)V^T \hat{w}_{\mathrm{NN}} - r^T d(t) + r^T \tilde{f} + r^T \tau_d + r^T v$$
$$+ \frac{1}{2}r^T \dot{M}r + \operatorname{tr}[\tilde{W}^T S^{-1}\dot{\tilde{W}}] + \operatorname{tr}[\tilde{W}_i^T T^{-1}\dot{\tilde{W}}_i].$$

Applying robot Property 3 and the tuning rules, one has

$$\dot{L} = -r^T K_V r + \operatorname{tr}[\tilde{W}^T (S^{-1}\dot{\tilde{W}}^T - \sigma'(V^T u + v_0)V^T \hat{w}_{\mathrm{NN}}r^T)]$$
$$+ \operatorname{tr}[\tilde{W}_i^T (T^{-1}\dot{\tilde{W}}_i^T + \sigma_i(V_i^T w + v_{0i})r^T \hat{W}^T \sigma'(V^T u + v_0)V^T)]$$
$$+ r^T(v - d(t) + \tilde{f} + \tau_d),$$

$$\dot{L} = -r^T K_V r + \text{tr}[\tilde{W}^T (S^{-1} \dot{\tilde{W}}^T - \sigma'(V^T u + v_0) V^T \hat{w}_{NN} r^T)]$$
$$+ \text{tr}[\tilde{W}_i^T (T^{-1} \dot{\tilde{W}}_i^T + \sigma_i (V_i^T w + v_{0i}) r^T \hat{W}^T \sigma'(V^T u + v_0) V^T)]$$
$$+ r^T (v - d(t) + \tilde{f} + \tau_d),$$
$$\dot{L} = -r^T K_V r + k_1 \|r\| \text{tr}[\tilde{W}^T (W - \tilde{W})]$$
$$+ \|r\| \text{tr}[\tilde{W}_i^T k_1 (W_i - \tilde{W}_i) + \tilde{W}_i^T k_2 \|\hat{W}_i\|_F (W_i - \tilde{W}_i)] - r^T d(t)$$
$$+ r^T (v + \tilde{f} + \tau_d).$$

Using the inequality

$$\text{tr}[\tilde{X}^T (X - \tilde{X})] \le \|\tilde{X}\|_F \|X\|_F - \|\tilde{X}\|_F^2$$

and Lemma 4.2.2, one may write

$$\dot{L} \le -K_{V\min} \|r\|^2 + k_1 \|r\| \|\tilde{W}\|_F (W_M - \|\tilde{W}\|_F)$$
$$+ k_1 \|r\| \|\tilde{W}_i\|_F (W_{iM} - \|\tilde{W}_i\|_F) + k_2 \|r\| \|\tilde{W}_i\|_F \|W_i - \tilde{W}_i\|_F (W_{iM} - \|\tilde{W}_i\|_F)$$
$$+ \|r\| (a_1 \|\tilde{W}\|_F + a_2 \|\tilde{W}_i\|_F^2 + a_3 \|\tilde{W}_i\|_F + a_5) - \|r\| (f_M + \tau_M) + \|r\| \|\tilde{f} + \tau_d\|,$$

$$\dot{L} \le -K_{V\min} \|r\|^2 + k_1 \|r\| \|\tilde{W}\|_F (W_M - \|\tilde{W}\|_F) + k_1 \|r\| \|\tilde{W}_i\|_F (W_{iM} - \|\tilde{W}_i\|_F)$$
$$+ k_2 \|r\| \|\tilde{W}_i\|_F \|W_i - \tilde{W}_i\|_F W_{iM} - k_2 \|r\| \|\tilde{W}_i\|_F^2 \|W_i - \tilde{W}_i\|_F$$
$$+ \|r\| (a_1 \|\tilde{W}\|_F + a_2 \|\tilde{W}_i\|_F^2 + a_3 \|\tilde{W}_i\|_F + a_5) - \|r\| (f_M + \tau_M) + \|r\| \|\tilde{f} + \tau_d\|$$

$$\le -K_{V\min} \|r\|^2 + k_1 \|r\| \|\tilde{W}\|_F (W_M - \|\tilde{W}\|_F) + k_1 \|r\| \|\tilde{W}_i\|_F (W_{iM} - \|\tilde{W}_i\|_F)$$
$$+ k_2 \|r\| \|\tilde{W}_i\|_F W_{iM}^2 + 2k_2 \|r\| \|\tilde{W}_i\|_F^2 W_{iM} - k_2 \|r\| \|\tilde{W}_i\|_F^3$$
$$+ \|r\| (a_1 \|\tilde{W}\|_F + a_2 \|\tilde{W}_i\|_F^2 + a_3 \|\tilde{W}_i\|_F + a_5) - \|r\| (f_M + \tau_M) + \|r\| \|\tilde{f} + \tau_d\|$$

$$\dot{L} \le -\|r\|\{K_{V\min} \|r\| - k_1 \|\tilde{W}\|_F (W_M - \|\tilde{W}\|_F) - k_1 \|\tilde{W}_i\|_F (W_{iM} - \|\tilde{W}_i\|_F)$$
$$- k_2 \|\tilde{W}_i\|_F W_{iM}^2 - 2k_2 \|\tilde{W}_i\|_F^2 W_{iM} + k_2 \|\tilde{W}_i\|_F^3$$
$$- a_1 \|\tilde{W}\|_F - a_2 \|\tilde{W}_i\|_F^2 - a_3 \|\tilde{W}_i\|_F - a_5 + f_M + \tau_M - \|\tilde{f} + \tau_d\|\}$$

$$\dot{L} \le -\|r\|\{K_{V\min} \|r\| + k_1 \|\tilde{W}\|_F^2 - (k_1 W_M + a_1) \|\tilde{W}\|_F + k_2 \|\tilde{W}_i\|_F^3$$
$$+ (k_1 - 2k_2 W_{iM} - a_2) \|\tilde{W}_i\|_F^2 - (k_1 W_{iM} + k_2 W_{iM}^2 + a_3) \|\tilde{W}_i\|_F$$
$$- a_5 + f_M + \tau_M - \|\tilde{f} + \tau_d\|\},$$

$$\dot{L} \le -\|r\|\left\{ K_{V\min} \|r\| + k_1 \left[\|\tilde{W}\|_F - \frac{1}{2} \left(W_M + \frac{a_1}{k_1} \right) \right]^2 \right.$$
$$\left. - k_1 \frac{1}{4} \left(W_M + \frac{a_1}{k_1} \right)^2 + g(\|\tilde{W}_i\|_F) - a_5 + f_M + \tau_M - \|\tilde{f} + \tau_d\| \right\},$$

where the function $g(x)$ is defined as

$$g(x) = k_2 x^3 + (k_1 - 2k_2 W_{iM} - a_2) x^2 - (k_1 W_{iM} + k_2 W_{iM}^2 + a_3) x.$$

Let the constant C be defined by

$$C = \inf\{g(x), x > 0\}.$$

Defining

$$h(x) = g(x) + C,$$

one has $h(x) \geq 0$ for every $x > 0$. Then one has

$$\dot{L} \leq -\|r\| \left\{ K_{V\,\min}\|r\| + k_1 \left[\|\tilde{W}\|_F - \frac{1}{2}\left(W_M + \frac{a_1}{k_1}\right) \right]^2 - k_1 \frac{1}{4}\left(W_M + \frac{a_1}{k_1}\right)^2 \right.$$

$$\left. + h(\|\tilde{W}_i\|_F) - C - a_5 - \|\tilde{f} + \tau_d\| + f_M + \tau_M \right\}.$$

Therefore, $\dot{L}$ is guaranteed negative as long as

$$\|r\| \geq \frac{\frac{1}{4}k_1\left(W_M + \frac{a_1}{k_1}\right)^2 + C + a_5}{K_{V\,\min}}$$

or

$$k_1\left[\|\tilde{W}\|_F - \frac{1}{2}\left(W_M + \frac{a_1}{k_1}\right)\right]^2 \geq \frac{1}{4}k_1\left(W_M + \frac{a_1}{k_1}\right)^2 + C + a_5$$

or

$$h(\|\tilde{W}_i\|_F) \geq \frac{1}{4}k_1\left(W_M + \frac{a_1}{k_1}\right)^2 + C + a_5.$$

The last three inequalities are equivalent to

$$\|r\| \geq \frac{\frac{1}{4}k_1\left(W_M + \frac{a_1}{k_1}\right)^2 + C + a_5}{K_{V\,\min}}, \tag{4.2.34}$$

$$\|\tilde{W}\|_F \geq \sqrt{\frac{1}{4}\left(W_M + \frac{a_1}{k_1}\right)^2 + \frac{C + a_5}{k_1}} + \frac{1}{2}\left(W_M + \frac{a_1}{k_1}\right), \tag{4.2.35}$$

$$\|\tilde{W}_i\|_F \geq \max\left\{ h^{-1}\left(\frac{1}{4}k_1\left(W_M + \frac{a_1}{k_1}\right)^2 + C + a_5\right) \right\}. \qquad \Box \tag{4.2.36}$$

We now remark on some special properties of the NN controller just described.

Coupled neural network weight-tuning laws

The mutual dependence between NN I and NN II results in the fact that the tuning laws (4.2.32) and (4.2.33) are coupled. This mathematical result followed from the fact that the pieces of information stored in NN I and NN II are dependent on each other due to (4.2.2). In fact, the proposed NN compensator with two NNs can be viewed as an NN compensator of *second order*. In effect, the two single-layer NNs function together as *a single NN with two tunable layers*. This is clearly seen from the tuning algorithms, where the gradient term in (4.2.33) is reminiscent of the gradient of the second layer of an NN that appears in tuning the first layer using backpropagation.

Modified backpropagation tuning

The first terms of (4.2.32) and (4.2.33) are modified versions of the standard backpropagation algorithm. The k_1 terms correspond to the e-mod (Narendra and Annaswamy (1987)), which is required to guarantee bounded NN weight estimates in the presence of the system disturbances. These terms are proportional to $k\|r\|\hat{W}$. The rationale for using such a term is that it tends to zero with the tracking error. Therefore, when there is no disturbance present, the correction term tends to zero as the tracking error approaches a zero value.

Note that the term corresponding to k_2 in (4.2.33) is a "second-order" e-mod, which is an efficient way to compensate for the second-order modeling mismatching term $a_2\|\tilde{W}_i\|_F^2$ in Lemma 4.2.2. Using these e-mod terms allows one to avoid persistence of excitation condition since the differential equations (4.2.32) and (4.2.33) are robust to system disturbances.

Bound on the tracking error

The right-hand side of (4.2.34) can be taken as a practical bound on the tracking error in the sense that $\|r(t)\|$ will never stray far above it. Note that the stability radius may be decreased to any desired value by increasing the PD gain K_v. It is noted that PD control without deadzone compensation requires much higher gains in order to achieve similar performance—that is, eliminating the NN feedforward compensator will result in degraded performance. Moreover, it is difficult to guarantee the stability of such a highly nonlinear system using only PD. Using the NN deadzone compensation, the stability of the system is proven, and the tracking error can be kept arbitrarily small by increasing the gain K_v.

According to (4.2.35) and (4.2.36), the NN weight errors are fundamentally bounded in terms of W_M. Thus one cannot guarantee they are small, so the NN weights may not converge to their real, ideal values. Still, the NN weights remain in the bounded neighborhood of the ideal weights, allowing one to prove system stability. As long as the tracking error is small, we do not care if the weights converge as long as they are bounded. The tuning parameters k_1, k_2 offer a design tradeoff between the relative eventual magnitudes of $\|r\|$ and $\|\tilde{W}\|_F$, $\|\tilde{W}_i\|_F$.

Initialization of neural network weights

Other NN control algorithms in the literature often rely on finding initial stabilizing NN weights, which is nearly impossible for practical systems.. By contrast, the weights in this NN controller are easy to initialize. The weights V and V_i are set to random values and left fixed. It is shown in Igelnik and Pao (1995) that for such NNs, termed random variable functional link (RVFL) NNs, the approximation property holds. The portion of the weight matrix V_i that corresponds to the jump function basis nodes is set to the values determined from the known location of the deadzone inverse jump at the origin (Chapter 3). The tuned weights W are initialized to random values, and W_i are initialized at zero. Then at the initial time, the NN II is effectively absent in Figure 4.2.5; there is only the unity feedforward path in the deadzone compensator. Therefore, the PD loop in Figure 4.2.5 holds the system stable until the NN begins to learn. Now one sees the importance of decomposing the deadzone inverse into a unity feedforward path plus an unknown jump function as performed in Figure 4.2.3.

No linearity in the model parameters

The proposed NN deadzone compensator does not require LIP. The standard techniques for deadzone compensation usually require such an assumption (Tao and Kokotović (1996),

Lewis et al. 1997)). The LIP requirement is a severe restriction for practical systems. We use here the NN approximation property, augmented for piecewise continuous functions, which holds for any jump function over a compact set. The nonlinear nonsymmetric deadzone model is linear with respect to the nonlinear NN activation functions, which is a fundamental difference from the LIP condition since the activation functions form a basis.

Since the LIP condition is not required, the above technique can be applied for inverting any unknown right-invertible function.

Adaptive critic architecture

The deadzone compensation scheme consists of two NNs: an NN estimator NN I and an NN compensator NN II. The NN compensator is placed in the feedforward loop, which is novel in the NN control literature, where NN compensators usually appear in the feedback loop. The NN estimator serves as a *performance evaluator* for the NN compensator and is used to tune it. This idea of one learning system supervising another is known as the *adaptive critic architecture*. The adaptive critic is usually an NN in a performance loop that appears outside or above all the feedback loops. In this particular application, the adaptive critic enjoys the unusual distinction of appearing inside the feedforward control loop.

4.2.4   Simulation of neural network deadzone compensator for robotic systems

To illustrate the performance of the NN deadzone compensator, the two-link robot arm in Figure 4.2.6 is used. The dynamics of this system are given in Lewis, Abdallah, and Dawson (1993). The system parameters are chosen as $\ell_1 = 1$, $\ell_2 = 1$, $m_1 = 1.8$, and $m_2 = 1.3$. The manipulator is acting in the horizontal plane, so there is no gravity effect. We simulated the two-link robot arm with deadzones in both links.

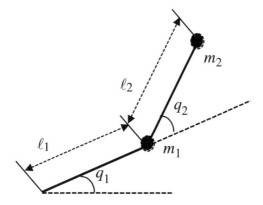

Figure 4.2.6. *Two-link robot arm.*

NN I is selected to have $L = 20$ hidden-layer nodes with sigmoidal activation functions. The first-layer weights V are selected randomly. They are uniformly randomly distributed between -1 and $+1$. These weights represent the stiffness of the sigmoidal activation function. The threshold weights for the first layer are uniformly randomly distributed between -35 and $+35$. The threshold weights represent the bias in the activation function positions. Therefore, they should cover the range of the deadzone. Since the deadzone width is not known, it is recommended that this range be selected large enough so that it entirely covers the deadzone width.

The tuning law requires that the second-layer weights $\hat{W}$ and $\hat{W}_i$ cannot both be initialized at zero because it is clear from (4.2.32) and (4.2.33) that in that case the NN weights would stay at zero forever. Therefore, the second-layer weights $\hat{W}$ for NN I are uniformly randomly initialized between -50 and $+50$. Note that this weight initialization will not affect system stability since the weights $\hat{W}_i$ are initialized at zero, and therefore there is initially no input to the system except for the PD loop, which keeps the trajectories bounded until the NN begins to learn and compensate for the deadzone.

NN II is augmented for the approximation of discontinuous functions as described in Chapter 3 and is chosen to have $L = 20$ hidden-layer nodes with sigmoidal activation functions plus four additional nodes with jump basis functions (two for each robot link). The first-layer weights V_i are uniformly randomly distributed between -1 and $+1$ as in NN I, and the threshold weights are uniformly randomly distributed between -35 and $+35$. The portion of the weight vector V_i corresponding to the jump basis functions are set to the values determined from the known location of the jump in the deadzone inverse. The second-layer weights $\hat{W}_i$ are initialized at zero.

The size L of the NNs should be selected to satisfy the system performance, but not too large. One way to do this is to select a smaller number L of the hidden-layer neurons and then increase the number in successive simulations until we get satisfactory performance of the system behavior. That is, at some point, the performance stops improving as we increase L.

To focus on the deadzone compensation, we selected the disturbance as $\tau_d(t) = 0$ and assumed that the controller knows the nonlinear robot function exactly, so that $\hat{f} = f$. This means that the robust term $v(t) = 0$. The NN weight-tuning parameters were chosen as $S = \text{diag}\{240, 240\}$, $T = \text{diag}\{500, 500\}$, $k_1 = 0.001$, and $k_2 = 0.0001$. The controller parameters are chosen as $\Lambda = \text{diag}\{7, 7\}$ and $K_v = \{15, 15\}$.

The deadzone is assumed to have linear functions outside the deadband. We selected $d_+ = 25$, $d_- = -20$, $h(u) = 1.5(u - d_+)$, and $g(u) = 0.8(u + d_-)$. We simulated two cases corresponding to two reference inputs.

Step reference input

First, a step signal is applied as the reference input to each robot joint. The position errors to unit step inputs for the first- and second-joint desired trajectories are shown in Figures 4.2.7 and 4.2.8 using only PD without deadzone compensation (using equation (4.2.28) with $\hat{w}_{NN} = 0$) and then using PD plus the NN compensator. The NN compensator signal $\hat{w}_{NN}(t)$ and the total control signal $u(t)$ are shown in Figures 4.2.9–4.2.12. One sees clearly that the tracking performance is dramatically improved by using the NN compensator.

Sinusoidal reference input

The same compensator is simulated when sinusoidal reference trajectory inputs are applied to each joint. The tracking errors for the first and second joints are shown in Figures 4.2.13 and 4.2.14 using only PD without NN deadzone compensation, and then using PD plus the NN compensator. The NN compensator signal $\hat{w}_{NN}(t)$ and the total control signal $u(t)$ are shown in Figures 4.2.15–4.2.18.

One can see that after a transient period of 1.5 seconds, the NNs adapt their weights in order to decrease the filtered tracking error $r(t)$. Thus the performance using the NN compensator is significantly better than the performance using only PD control.

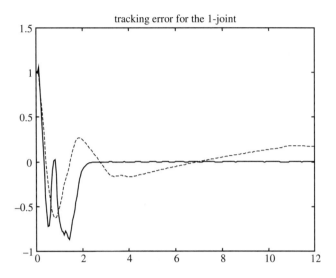

Figure 4.2.7. *Position error for 1-joint: Without deadzone compensation (dash) and with NN deadzone compensator (full).*

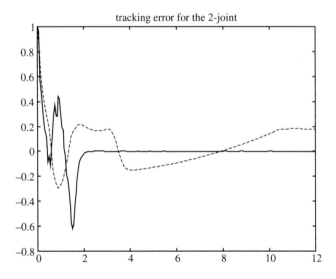

Figure 4.2.8. *Position error for 2-joint: Without deadzone compensation (dash) and with NN deadzone compensator (full).*

Adaptive dithering

It is interesting to note that the NN compensator signal $\hat{w}_{NN}(t)$ has a superimposed high-frequency component that is very similar to that injected using dithering techniques (Desoer and Shahruz (1986)). In fact, the signal from the NN is injected at the same position in the control loop where dithering signals are injected. Therefore, one could consider our NN deadzone compensator as an *adaptive dithering technique*.

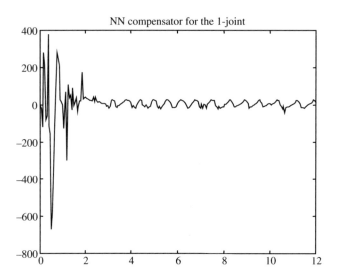

Figure 4.2.9. *NN compensator signal for the first joint.*

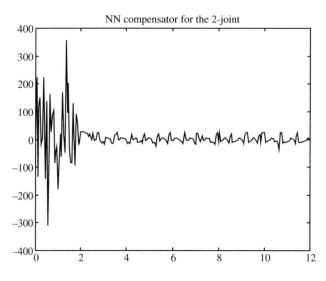

Figure 4.2.10. *NN compensator signal for the second joint.*

4.3   Fuzzy logic discrete-time deadzone precompensation

Many systems with deadzones are modeled in discrete time. An example in biomedical control is the functional neuromuscular stimulation for restoring motor function by directly activating paralyzed muscles (Bernotas, Crago, and Chizeck (1987)). Moreover, for implementation in digital controllers, a discrete-time deadzone compensator is needed. To address discrete-time deadzone compensation, an adaptive control approach has been proposed in Tao and Kokotović (1995). There, a projection algorithm was used to estimate the deadzone parameters. In keeping with standard adaptive control techniques applied to discrete-time systems, it was assumed that the deadzone was LIP, and a certainty equivalence (CE) assumption was then made to prove stability.

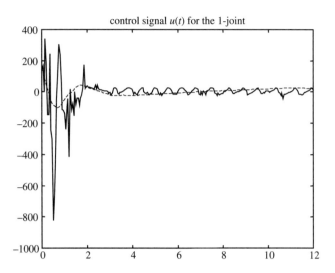

Figure 4.2.11. *Control signal $u(t)$ for the first joint: Without deadzone compensation (dash) and with NN deadzone compensator (full).*

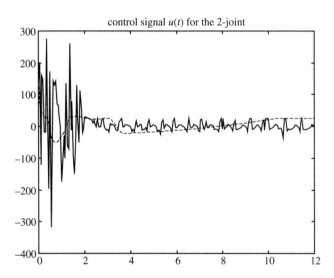

Figure 4.2.12. *Control signal $u(t)$ for the second joint: Without deadzone compensation (dash) and with NN deadzone compensator (full).*

The use of FL systems has accelerated in recent years in many areas, including feedback control. Particularly important in FL control are the *universal function approximation capabilities* of FL systems discussed in Chapter 1. Given these recent results, some rigorous design techniques for FL feedback control based on *adaptive control* approaches have been given (Wang (1997)). FL systems offer significant advantages over adaptive control, including no requirement for LIP and no need to compute a regression matrix for each specific system. Actuator nonlinearities are typically defined in terms of piecewise linear functions according to the region to which the argument belongs. The FL function approximation properties and the ability of FL systems to discriminate information based on regions of the

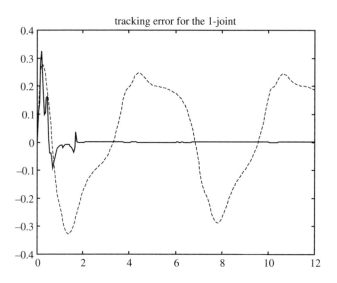

Figure 4.2.13. *Tracking error for 1-joint: Without deadzone compensation* (*dash*) *and with NN deadzone compensator* (*full*).

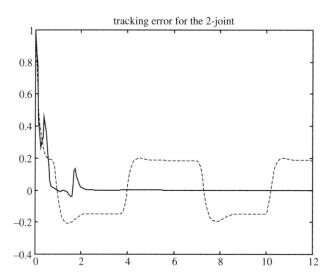

Figure 4.2.14. *Tracking error for 2-joint: Without deadzone compensation* (*dash*) *and with NN deadzone compensator* (*full*).

input variables (the classification property) make them an ideal candidate for compensation of nonanalytic actuator nonlinearities. FL deadzone compensation schemes are provided in Kim et al. (1993, 1994) and Lewis et al. (1997).

Here we show how to design a motion tracking controller for discrete-time multi-input Lagrangian mechanical systems with unknown input deadzones. The general case of nonsymmetric deadzones is treated. A rigorous design procedure is given that results in a PD tracking loop with an adaptive FL system in the feedforward loop for deadzone compensation. The FL feedforward deadzone compensator is adapted in such a way as

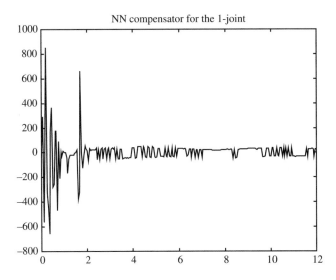

Figure 4.2.15. *NN compensator signal for the first joint.*

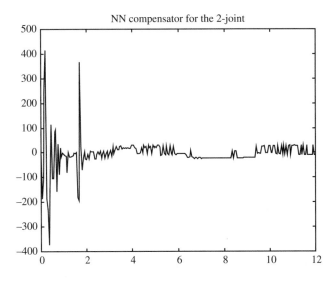

Figure 4.2.16. *NN compensator signal for the second joint.*

to estimate online the unknown width of the deadzones. Unlike standard discrete-time adaptive control techniques, no CE assumption is needed since both the tracking error and the estimation error are weighted *in the same* Lyapunov function. The approach is similar to that in Jagannathan and Lewis (1996), but additional complexities arise due to the fact that the FL system is in the *feedforward* loop.

4.3.1   Fuzzy logic discrete-time deadzone precompensator

In this subsection, an FL discrete-time precompensator is designed for nonsymmetric dead-zone nonlinearities in the actuation of industrial processes. In the next subsection we show

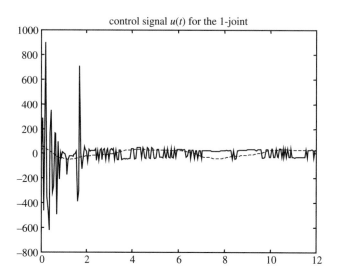

Figure 4.2.17. *Control signal u(t) for the first joint: Without deadzone compensation (dash) and with NN deadzone compensator (full).*

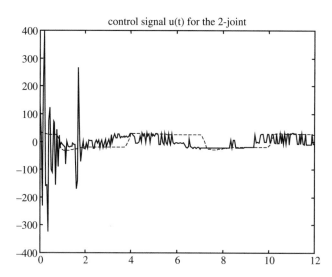

Figure 4.2.18. *Control signal u(t) for the second joint: Without deadzone compensation (dash) and with NN deadzone compensator (full).*

how to use this FL precompensator to build a tracking controller for industrial processes with deadzones modeled in discrete time.

If u and τ are scalars, the nonsymmetric deadzone nonlinearity for a discrete-time system, shown in Figure 4.3.1, is given by

$$\tau(k) = D_d(u(k)) = \begin{cases} u(k) + d(k)_-, & u(k) < -d(k)_-, \\ 0, & -d(k)_- \leq u(k) < d(k)_+, \\ u(k) - d(k)_+, & d(k)_+ \leq u(k). \end{cases} \qquad (4.3.1)$$

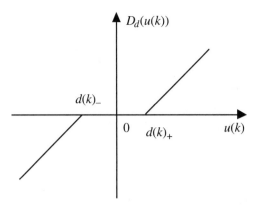

Figure 4.3.1. *Discrete-time nonsymmetric deadzone nonlinearity.*

The parameter vector $d(k) = [d(k)_+ \quad d(k)_-]^T$ characterizes the width of the motion deadband. In practical motion-control systems, the width of the deadzone is unknown, so compensation is difficult. Most compensation schemes cover only the case of symmetric deadzones where $d(k)_- = d(k)_+$.

The nonsymmetric deadzone may be written as

$$\tau(k) = D_d(u(k)) = u(k) - \text{sat}_d(u(k)), \tag{4.3.2}$$

where the nonsymmetric saturation function is defined as

$$\text{sat}_d(u(k)) = \begin{cases} -d(k)_-, & u(k) < -d(k)_-, \\ u(k), & -d(k)_- \le u(k) < d(k)_+, \\ d(k)_+, & d(k)_+ \le u(k). \end{cases} \tag{4.3.3}$$

This is a decomposition into a unity feedforward path plus an unknown parallel path. This decomposition is vital for the success of the control scheme to be derived.

To offset the deleterious effects of deadzone, one may place a precompensator as illustrated in Figure 4.3.2. There the desired function of the precompensator is to cause the composite throughput from w to τ to be unity. The power of FL systems is that they allow one to use intuition, based on experience, to design control systems and then provide the mathematical machinery for rigorous analysis and modification of the intuitive knowledge, for example, through learning or adaptation, to give guaranteed performance. Due to the FL classification property, they are particularly powerful when the nonlinearity depends on the region in which the argument u of the nonlinearity is located, as in the nonsymmetric deadzone.

In most practical motion control systems, there are several control inputs so that w, u, and τ are generally n-vectors. Then there may be different deadzone widths in each channel so that for $i = 1, \ldots, n$, for each component w_i, u_i, τ_i, one has the nonsymmetric deadzone

$$\tau_i = D_{di}(u_i) = u_i - \text{sat}_{di}(u_i) \tag{4.3.4}$$

with $d_i = [d_{i+} \quad d_{i-}]^T$. One can write this in vector form

$$\tau_i = D_D(u) = u - \text{sat}_D(u), \tag{4.3.5}$$

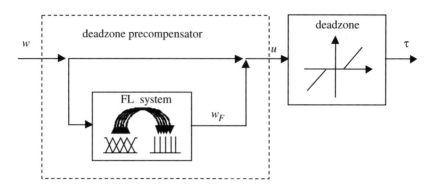

Figure 4.3.2. *FL deadzone compensation scheme.*

where the block diagonal matrix of deadzone widths is $D \equiv \mathrm{diag}\{d_1, d_2, \ldots, d_n\} \in R^{2n \times n}$. The vector saturation function is defined as

$$\mathrm{sat}_D(u) \equiv [\mathrm{sat}_{di}(u_i)], \qquad (4.3.6)$$

where $[z_i]$ denotes the vector with component z_i. Then one must use an FL compensator in each channel.

A deadzone precompensator, designed using engineering experience, would be discontinuous and depend on the region within which w occurs. It would be naturally described using the rules

$$\text{if } w_i \in X_+(w_i), \quad \text{then } w_{Fi} = \hat{d}_{i+};$$
$$\text{if } w_i \in X_-(w_i), \quad \text{then } w_{Fi} = -\hat{d}_{i-}.$$

To accomplish this in a rigorous setting, we introduce an FL framework, defining

$$u_i = w_i + w_{Fi} \qquad (4.3.7)$$

with membership functions (MFs) $X_+(\cdot)$ and $X_-(\cdot)$ defined for each component w_i according to the following:

$$X_+(w_i) = \begin{cases} 0, & w_i < 0, \\ 1, & 0 \le w_i, \end{cases}$$
$$\qquad\qquad\qquad\qquad (4.3.8)$$
$$X_-(w_i) = \begin{cases} 1, & w_i < 0, \\ 0, & 0 \le w_i. \end{cases}$$

These MFs are shown in Figure 4.3.3.

Define the estimate vector $\hat{d}_i = [\hat{d}_{i+} \quad \hat{d}_{i-}]^T$. The FL precompensator may be conveniently expressed in vector form as follows. Define the vector $w_F = [w_{F1} \quad w_{F2} \quad \cdots \quad w_{Fn}]^T$ so that

$$u = w + w_F,$$
$$\qquad\qquad\qquad\qquad (4.3.9)$$
$$u = w + \hat{D}^T X(w),$$

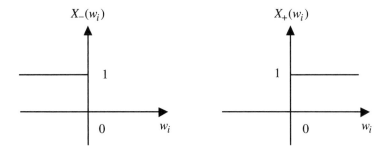

Figure 4.3.3. *MFs for FL deadzone precompensator.*

where the block diagonal matrix of estimated deadzone widths is $\hat{D} \equiv \mathrm{diag}\{\hat{d}_1, \hat{d}_2, \ldots, \hat{d}_n\}$ and the vector FL basis function is given by

$$
X(w) = \begin{bmatrix} X_+(w_1) \\ -X_-(w_1) \\ \vdots \\ X_+(w_n) \\ -X_-(w_n) \end{bmatrix}.
\tag{4.3.10}
$$

As shown in Lewis et al. (1997), the throughput of the compensator plus deadzone for vectors $w, u, \tau \in R^n$ is

$$
\tau = w - \tilde{D}^T X(w) + \tilde{D}^T \delta,
\tag{4.3.11}
$$

where the matrix deadzone width estimation is $\tilde{D} = D - \hat{D} \equiv \mathrm{diag}\{\bar{d}_1, \bar{d}_2, \ldots, \bar{d}_n\}$. The vector mismatch δ is bounded by

$$
\|\delta\| \le \sqrt{n}.
\tag{4.3.12}
$$

This result corresponds to that in Theorem 4.2.1 for NN precompensation of deadzone. Its form is very important in the upcoming controller design phase. Note that it says that the throughput is almost equal to the identify plus some small terms coming from the deadzone parameter estimation errors and a bounded disturbance.

4.3.2 Discrete-time fuzzy logic controller with deadzone compensation

Dynamics of an mnth-order multi-input, multi-output system

Many industrial processes may be modeled in discrete-time by the mnth-order multi-input, multi-output (MIMO) nonlinear discrete-time system given by

$$
x_1(k+1) = x_2(k),
$$

$$
\vdots
$$

$$
x_{n-1}(k+1) = x_n(k),
$$

$$
x_n(k+1) = f(x(k)) + \tau(k) + \tau_d(k),
\tag{4.3.13}
$$

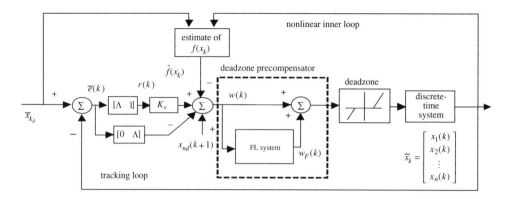

Figure 4.3.4. *Tracking controller with FL adaptive deadzone compensation.*

where $x(k) = [x_1(k), x_2(k), \ldots, x_n(k)]^T$ with $x_i(k) \in R^m$, $i = 1, 2, \ldots, n$, where $\tau(k) \in R^m$ denotes the control input acting on the system at the instant k. Signal $\tau_d(k) \in R^m$ denotes a disturbance vector acting on the system at the instant k, with $\|\tau_d(k)\| \leq d_M$ a known constant.

Given a desired motion trajectory $x_{nd}(k)$ and its delayed values, define the tracking error as

$$e_n(k) = x_n(k) - x_{nd}(k). \tag{4.3.14}$$

It is typical in robotics to define a so-called filtered tracking error, $r(k) \in R^m$, given by

$$r(k) = e_n(k) + \lambda_1 e_{n-1}(k) + \cdots + \lambda_{n-1} e_1(k), \tag{4.3.15}$$

where $e_{n-1}(k), \ldots, e_1(k)$ are the delayed values of the error $e_n(k)$ and $\lambda_1, \ldots, \lambda_{n-1}$ are constant matrices selected so that $|z^{n-1} + \lambda_1 z^{n-2} + \cdots + \lambda_{n-1}|$ is stable. Then one has

$$r(k + 1) = e_n(k + 1) + \lambda_1 e_{n-1}(k + 1) + \cdots + \lambda_{n-1} e_1(k + 1). \tag{4.3.16}$$

Define $\Lambda = [\lambda_{n-1} \quad \cdots \quad \lambda_1]$.

Using equation (4.3.13) in (4.3.16), the dynamics of the mnth-order MIMO system can be written in terms of the tracking error as

$$r(k + 1) = f(x(k)) - x_{nd}(k + 1) + \lambda_1 e_n(k) + \cdots + \lambda_{n-1} e_2(k) + \tau(k) + \tau_d(k). \tag{4.3.17}$$

In this section, it is shown how to tune or learn the deadzone width estimates $\hat{D}(k)$ in the control (4.3.9) online so that the tracking error is guaranteed small and all internal states are bounded. This turns the discrete-time deadzone compensator into a *discrete-time adaptive* FL deadzone compensator. It is assumed, of course, that the actuator output $\tau(k)$ is not measurable. The controller derived in the upcoming development is shown in Figure 4.3.4. It is important to note that the discrete-time FL deadzone compensation signal $w(k)$ is injected at exactly the same point at which standard dithering signals are injected (Desoer and Shahruz (1986)). Thus this deadzone compensation scheme could be considered as *adaptive dithering*.

In the upcoming proof, the following standard assumptions are made.

Assumption 5 (Bounded disturbances). The unknown disturbance satisfies $\|\tau_d(k)\| < \tau_M$, with τ_M a known positive constant.

Assumption 6 (Bounded nonlinear functional estimation error). The unknown nonlinear term satisfies $\|\tilde{f}(x(k))\| < f_M(x(k))$, with $f_M(x(k))$ a known bounding function.

Assumption 7 (Constant bounded deadzone). It is assumed that there is a known bound on the magnitude of the unknown deadzone widths so that $\|D(k)\| \le D_M$ for some known scalar D_M.

Moreover, it is assumed that the deadzone widths are constant, $D(k+1) = D(k)$.

It is desired to design a motion controller that causes the mechanical system to track a prescribed trajectory $x_d(k)$. To accomplish this, we defined the tracking error by equation (4.3.14) and the filtered error by equation (4.3.15). Working in terms of the filtered tracking error significantly simplifies the controller design stage since we may deal with the filtered error system (4.3.17) instead of the complete dynamics (4.3.13). Note that equation (4.3.15), with $r(k)$ considered as the input, is a stable system. This means that $e_n(k)$ is bounded as long as the controller guarantees that the filtered error $r(k)$ is bounded. Therefore, we are justified in designing a controller so that the filtered tracking error is bounded, for this guarantees stable motion tracking.

Tracking controller with fuzzy logic deadzone compensation

A compensation scheme for unknown terms in $f(x(k))$ is provided by selecting the tracking controller

$$w(k) = K_v \cdot r(k) - \hat{f}(x(k)) + x_{nd}(k+1) - \lambda_1 \cdot e_n(k) - \lambda_2 \cdot e_{n-1}(k) - \cdots - \lambda_{n-1} \cdot e_2(k) \tag{4.3.18}$$

with $\hat{f}(x(k))$ an estimate for the nonlinear terms $f(x(k))$. The feedback gain matrix $K_v > 0$ is often selected to be diagonal.

If the nonlinear function $f(x(k))$ is unknown, it can be estimated using adaptive control techniques or the FL controller in Wang (1997). In this development, we are interested in focusing on FL deadzone compensation, so we shall take a robust control approach (Corless and Leitmann (1982)) and assume that a fixed estimate $\hat{f}(x(k))$ is available. The estimate $\hat{f}(x(k))$ is computed by an inner nonlinear feedback linearization control loop.

It is assumed that the functional estimation error satisfies $\|\tilde{f}(x(k))\| \le f_M(x(k))$ for some known bounding function $f_M(x(k))$. This standard assumption is not unreasonable, since in practical systems the bound $f_M(x(k))$ can be computed knowing upper bounds on payload masses, frictional effects, and so on.

Deadzone compensation is provided using

$$u(k) = w(k) + \hat{D}^T(k) \cdot X(w(k)), \tag{4.3.19}$$

where $X(w(k))$ is given by (4.3.10), which gives the overall feedforward throughput (4.3.11). The multiloop control structure implied by this scheme is shown in Figure 4.3.4. The controller has a PD tracking loop, and the deadzone effect is ameliorated by a feedforward FL compensator.

To design the FL system so that the tracking error $r(k)$ is bounded and all internal states are stable, one must examine the error dynamics. Substituting (4.3.11) and (4.3.18) into (4.3.17) yields the closed-loop error dynamics

$$r(k+1) = K_v \cdot r(k) - \tilde{D}^T(k) \cdot X(w(k)) + \tilde{D}^T(k) \cdot \delta + \tilde{f}(x(k)) + \tau_d(k), \tag{4.3.20}$$

where the nonlinear functional estimation error is given by $\tilde{f}(x(k)) = f(x(k)) - \hat{f}(x(k))$.

The next theorem is our main result; it provides an algorithm for tuning the FL discrete-time deadzone precompensator. It guarantees stability in the sense of uniform ultimate boundedness (see Chapter 2).

Theorem 4.3.1 (Controller with FL deadzone compensation). *Consider the system given by (4.3.13). Let the desired trajectory $x_{nd}(k)$ and the initial conditions be bounded in a compact set S. Let the disturbance and the deadzone matrix bounds τ_M and D_M, respectively, be known constants.*

Let $u(k)$ be given by (4.3.19) and the FL parameter tuning law be provided by

$$\hat{D}(k+1) = \hat{D}(k) - 2\alpha X(w(k))r^T(k+1) - \Gamma\|I - 4\alpha X(w(k))X^T(w(k))\|\hat{D}(k),$$
$$(4.3.21)$$

where $\alpha > 0$ is a constant learning rate parameter or adaptation gain and $\Gamma > 0$ is a design parameter. Then the filtered tracking error $r(k)$ and the deadzone estimates $\hat{D}(k)$ are UUB, provided the following conditions hold:

$$(1) \quad \|X\|^2 < \frac{1}{(4\cdot\alpha)}, \tag{4.3.22}$$

$$(2) \quad \frac{1}{4} < \Gamma < 1, \tag{4.3.23}$$

$$(3) \quad K_{v\,\max} < \sqrt{\frac{\rho_1}{\bar{\sigma}(\rho_1 + \bar{\sigma}\|X\|^2)}}, \tag{4.3.24}$$

where

$$\bar{\sigma} = 1 + 4\alpha X^T X + \frac{[2\Gamma\|I - 4\alpha XX^T\| - (1 - 4\alpha X^T X)]^2}{1 - 4\alpha X^T X} \tag{4.3.25}$$

and

$$\rho_1 = \frac{\|I - 4\alpha XX^T\|^2}{\alpha}\Gamma(2 - \Gamma) + 4\Gamma\|I - 4\alpha XX^T\|\|X\|^2$$
$$- \frac{[1 - 4\alpha X^T X - 2\Gamma\|I - 4\alpha XX^T\|]^2}{1 - 4\alpha X^T X} \tag{4.3.26}$$
$$> 0.$$

Note that $\bar{\sigma} > 0$ because of condition (4.3.22).

Proof of condition (4.3.26).

$$\rho_1 = \frac{\|I - 4\alpha XX^T\|^2}{\alpha}\Gamma(2 - \Gamma) + 4\Gamma\|I - 4\alpha XX^T\|\cdot\|X\|^2$$
$$- \frac{(1 - 4\alpha X^T X - 2\Gamma\|I - 4\alpha\cdot X^T\|)^2}{1 - 4\alpha X^T X}\|X\|^2$$
$$> 0.$$

Using 2-norm bounds properties and condition (4.3.22),

$$\rho_1 > 4\|X\|^2(1 - 4\alpha \cdot X^T X)^2 \Gamma(2 - \Gamma) + 8 \cdot \Gamma(1 - 4\alpha \cdot X^T X) \cdot \|X\|^2$$
$$- (1 - 4\alpha \cdot X^T X)\|X\|^2 - 4\Gamma^2(1 - 4\alpha \cdot X^T X)\|X\|^2$$
$$= \|X\|^2(1 - 4\alpha \cdot X^T X)[4(1 - 4\alpha \cdot X^T X)\Gamma(2 - \Gamma) + 4\Gamma - 1]$$
$$+ 4\Gamma(1 - \Gamma)(1 - 4\alpha \cdot X^T X)\|X\|^2.$$

Using conditions (4.3.22) and (4.3.23), it can be seen that $\rho_1 > 0$.   □

***Proof of Theorem* 4.3.1.** For the proof, see the appendix of this chapter (section 4.A).

It is important to note that in this theorem there is no CE assumption, in contrast to standard work in discrete-time adaptive control. In the latter, a parameter identifier is first selected and the parameter estimation errors are assumed to be small. Then in the tracking proof, it is assumed that the parameter estimates are exact (the CE assumption), and a Lyapunov function is selected that weights only the tracking error to demonstrate closed-loop stability and tracking performance. This approach is used, for instance, in Tao and Kokotović (1995).

By contrast, in our proof, the Lyapunov function in the appendix of this chapter (section 4.A) is of the form

$$J = r^T \cdot r + \frac{1}{\alpha} \text{tr}\{\tilde{D}^T \cdot \tilde{D}\}, \tag{4.3.27}$$

which weights *both* the tracking error $r(k)$ and the parameter estimation error $\tilde{D}(k)$. This is highly unusual in discrete-time control system design. The problem with weighting both the tracking error and the parameter estimation error is that in discrete time, the Lyapunov function difference is quadratic, which means it is very complicated to handle. It requires an exceedingly complex proof (note that we are forced to complete the square several times with respect to different variables), but it obviates the need for any sort of CE assumption. It also allows the parameter-tuning algorithm to be derived during the proof process, not selected a priori in an ad hoc manner. This is akin to the proof of Jagannathan and Lewis (1996), but additional complexities arise due to the fact that the FL system is in the *feedforward* loop.

The third term in (4.3.21) is a discrete-time version of Narendra's e-mod (Narendra and Annaswamy (1987)), which is required to provide robustness due to the coupling in the proof between tracking error terms and parameter error terms in the Lyapunov function. It is similar to what have been called "forgetting terms" in NN weight-tuning algorithms. These are required in that context to prevent parameter overtraining.

4.3.3   Simulation results

In this section, the discrete-time FL compensator is simulated on a digital computer. It is found to be very efficient at canceling the deleterious effects of actuator deadzones.

We simulate the response for the known plant with input deadzone, both with and without the FL compensator. The following nonlinear plant is used:

$$x_1(k + 1) = x_2(k),$$
$$x_2(k + 1) = -\frac{3}{16}\left[\frac{x_1(k)}{1 + x_2^2(k)}\right] + x_2(k) + u(k). \tag{4.3.28}$$

The deadzone widths were selected as $d_+ = d_- = 0.1$ for the first experiment and then $d_+ = d_- = 0.3$ for the second one.

We simulate the trajectory tracking performance of the system for sinusoidal desired signals. The desired motion trajectory used was selected to be

$$x_d(k) = \sin(w \cdot t_k + \phi)v, \quad w = 0.5, \quad \phi = \frac{\pi}{2}. \tag{4.3.29}$$

The initial estimates for the deadzone widths were selected as $d_+ = d_- = 0$ and the sampling period as $T = 100ms$.

Figure 4.3.5 shows the system response without deadzone using a standard PD controller. The PD controller does a good job on the tracking but a small steady-state tracking error remains. Figure 4.3.6 shows the system response with a deadzone having a width of $d_+ = d_- = 0.1$. The system deadzone destroys the tracking, and the PD controller is not capable of compensating for that.

Now an FL deadzone compensator was added. Figure 4.3.7 shows the same situation but using the proposed discrete-time FL deadzone compensator. The deadzone compensator

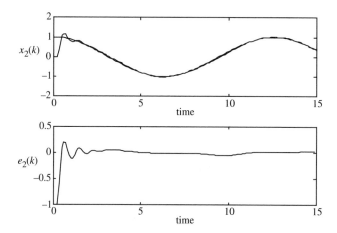

Figure 4.3.5. *PD controller without deadzone.*

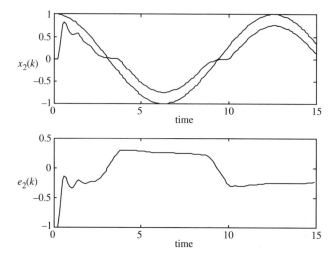

Figure 4.3.6. *PD controller with deadzone ($d = 0.1$).*

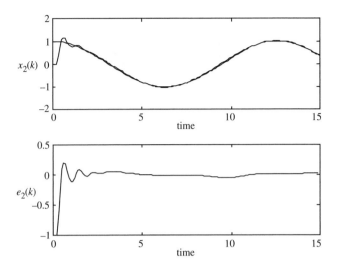

Figure 4.3.7. *PD controller with FL deadzone compensation* $(d = 0.1)$.

takes care of the system deadzone and the tracking is as good as it was without deadzone. The same result can be observed for a bigger deadzone value $d_+ = d_- = 0.3$ in Figures 4.3.8 and 4.3.9. The tracking performance is excellent even with large deadzones.

4.A Appendix: Proof of Theorem 4.3.1

For simplicity, let us rewrite the tracking error system given by (4.3.20) as

$$r_{k+1} = K_{vk}r_k - \tilde{D}_k^T X_k + \tilde{D}_k^T \delta + d_k, \qquad (4.A.1)$$

where $d_k = \tilde{f}_k + \tau_{dk}$.

Note: From now on, we will omit the k subindex for simplicity purposes. Thus every variable is suppose to have a k subindex unless otherwise specified.

Define the Lyapunov function candidate

$$J = r^T r + \frac{1}{\alpha} \text{tr}\{\tilde{D}^T \tilde{D}\}. \qquad (4.A.2)$$

The first difference is

$$\Delta J = \Delta J_1 + \Delta J_2 = r_{k+1}^T r_{r+1} - r^T r + \frac{1}{\alpha} \text{tr}\{\tilde{D}_{k+1}^T \tilde{D}_{k+1} - \tilde{D}^T \tilde{D}\}. \qquad (4.A.3)$$

Considering the first term from (4.A.3) and substituting (4.A.1),

$$\begin{aligned}
\Delta J_1 &= r_{k+1}^T r_{k+1} - r^T r \\
&= (K_v \cdot r - \tilde{D}^T X + \tilde{D}^T \delta + d)^T (K_v \cdot r - \tilde{D}^T X + \tilde{D}^T \delta + d) - r^T r \\
&= r^T \cdot K_v^T K_v \cdot r - 2 \cdot r^T K_v^T \tilde{D}^T X + 2 \cdot r^T K_v^T \tilde{D}^T \delta + 2 \cdot r^T K_v^T d + X^T \tilde{D} \tilde{D}^T X \\
&\quad - 2 \cdot X^T \tilde{D} \tilde{D}^T \delta - 2 \cdot X^T \tilde{D} \cdot d + \delta^T \tilde{D} \tilde{D}^T \delta + 2 \cdot \delta^T \tilde{D} \cdot d + d^T d - r^T r \\
&= r^T \cdot K_v^T K_v \cdot r - 2 \cdot r^T K_v^T \tilde{D}^T X + 2 \cdot r^T K_v^T \tilde{D}^T \delta + 2 \cdot r^T K_v^T d + X^T \tilde{D} \tilde{D}^T X \\
&\quad - 2 \cdot X^T \tilde{D} \tilde{D}^T \delta - 2 \cdot X^T \tilde{D} \cdot d + \delta^T \tilde{D} \tilde{D}^T \delta + 2 \cdot \delta^T \tilde{D} \cdot d + d^T d.
\end{aligned}$$

$$(4.A.4)$$

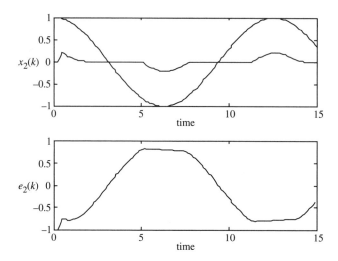

Figure 4.3.8. *PD controller with deadzone* $(d = 0.3)$.

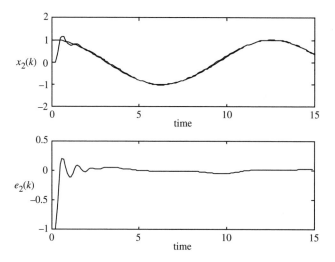

Figure 4.3.9. *PD controller with FL deadzone compensation* $(d = 0.3)$.

Now taking the second term in (4.A.3),

$$\Delta J_2 = \frac{1}{\alpha} \operatorname{tr}\{\tilde{D}_{k+1}^T \tilde{D}_{k+1} - \tilde{D}^T \tilde{D}\} = \frac{1}{\alpha} \operatorname{tr}\{(D_{k+1} - \hat{D}_{k+1})^T (D_{k+1} - \hat{D}_{k+1}) - \tilde{D}^T \tilde{D}\}$$

$$= \frac{1}{\alpha} \operatorname{tr}\{D_{k+1}^T D_{k+1} - 2 \cdot D_{k+1}^T \hat{D}_{k+1} + \hat{D}_{k+1}^T \hat{D}_{k+1} - \tilde{D}^T \tilde{D}\}.$$

Using the fact that the deadzone width is constant (i.e., $D_{k+1} = D_k$),

$$\Delta J_2 = \frac{1}{\alpha} \cdot \operatorname{tr}\{D^T D - 2 \cdot D^T \hat{D}_{k+1} + \hat{D}_{k+1}^T \hat{D}_{k+1} - \tilde{D}^T \tilde{D}\}.$$

Using the tuning law and simplifying,

$$
\begin{aligned}
\Delta J_2 &= \frac{1}{\alpha} \operatorname{tr}\{ D^T D - 2 \cdot D^T \{ \hat{D} - 2\alpha \cdot X r_{k+1}^T - \Gamma \| I - 4\alpha \cdot X X^T \| \hat{D} \} \\
&\quad + (\hat{D} - 2\alpha X \cdot r_{k+1}^T - \Gamma \| I - 4\alpha \cdot X X^T \| \hat{D})^T \\
&\quad \times (\hat{D} - 2\alpha X \cdot r_{k+1}^T - \Gamma \| I - 4\alpha \cdot X X^T \| \hat{D}) - \tilde{D}^T \tilde{D} \} \\
&= \frac{1}{\alpha} \operatorname{tr}\{ 4\alpha \cdot \tilde{D}^T X \cdot r_{k+1}^T + 4\alpha^2 \cdot r_{k+1} X^T X \cdot r_{k+1}^T + 2 \cdot \tilde{D}^T \Gamma \cdot \| I - 4\alpha \cdot X X^T \| \hat{D} \\
&\quad + 4\alpha \cdot r_{k+1} X^T \Gamma \| I - 4\alpha \cdot X X^T \| \hat{D} + \hat{D}^T \| I - 4\alpha \cdot X X^T \|^2 \Gamma^T \Gamma \hat{D} \}.
\end{aligned}
$$

Given a vector $x \in R^n$, $tr(x x^T) = x^T x$, using this above, we will obtain

$$
\begin{aligned}
\Delta J_2 &= 4 \cdot X^T \tilde{D} \cdot r_{k+1} + 4 \cdot \alpha \cdot X^T X \cdot r_{k+1}^T \cdot r_{k+1} + 4 \cdot r_{k+1}^T \hat{D}^T \Gamma^T X \| I - 4\alpha \cdot X X^T \| \\
&\quad + \frac{1}{\alpha} \operatorname{tr}\{ 2 \cdot \tilde{D}^T \Gamma \| I - 4\alpha \cdot X X^T \| \hat{D} + \| I - 4\alpha \cdot X X^T \|^2 \hat{D}^T \Gamma^T \Gamma \hat{D} \}.
\end{aligned}
$$

Substituting (4.A.1) in the above equation and putting the expression

$$
-8 X^T \tilde{D} \hat{D}^T \Gamma^T X \| I - \alpha X X^T \|
$$

back on the trace term,

$$
\begin{aligned}
\Delta J_2 &= 4 \cdot X^T \tilde{D} \cdot K_v \cdot r - 4 \cdot X^T \tilde{D} \tilde{D}^T X + 4 \cdot X^T \tilde{D} \tilde{D}^T \delta + 4 \cdot X^T \tilde{D} \cdot d \\
&\quad + 4 \cdot \alpha \cdot (X^T X) \cdot r^T K_v^T K_v \cdot r - 8\alpha \cdot (X^T X) \cdot r^T K_v^T \tilde{D}^T X \\
&\quad + 8\alpha \cdot (X^T X) \cdot r^T K_v^T \tilde{D}^T \delta + 8\alpha \cdot (X^T X) \cdot r^T K_v^T d \\
&\quad + 4\alpha \cdot (X^T X) X^T \tilde{D} \tilde{D}^T X - 8\alpha \cdot (X^T X) X^T \tilde{D} \tilde{D}^T \delta - 8\alpha \cdot (X^T X) X^T \tilde{D} \cdot d \\
&\quad + 4\alpha \cdot (X^T X) \delta^T \tilde{D} \tilde{D}^T \delta + 8\alpha \cdot (X^T X) \delta^T \tilde{D} \cdot d + 4 \cdot \alpha \cdot (X^T X) d^T d \\
&\quad + 4 r^T K_v^T \hat{D}^T \Gamma^T X \| I - 4\alpha X X^T \| + 4 X^T \tilde{D} \hat{D}^T \Gamma^T X \| I - 4\alpha \cdot X X^T \| \\
&\quad + 4 \delta^T \tilde{D} \hat{D}^T \Gamma^T X \| I - 4\alpha \cdot X X^T \| + 4 d^T \hat{D}^T \Gamma^T X \| I - 4\alpha \cdot X X^T \| \\
&\quad + \frac{1}{\alpha} \operatorname{tr}\{ 2 \cdot \tilde{D}^T \Gamma \| I - 4\alpha \cdot X X^T \| \hat{D} + \| I - \alpha \cdot X X^T \|^2 \hat{D}^T \Gamma^T \Gamma \hat{D} \\
&\quad - 8\alpha \cdot \tilde{D}^T X X^T \Gamma^T \| I - 4\alpha \cdot X X^T \| \hat{D} \}.
\end{aligned}
\tag{4.A.5}
$$

Substituting (4.A.4) and (4.A.5) in (4.A.2), collecting terms together, and using $\hat{D} = D - \tilde{D}$, we will have

$$
\begin{aligned}
\Delta J &= -r^T [I - (1 + 4\alpha \cdot X^T X) K_v^T K_v] \cdot r - (3 - 4\alpha \cdot X^T X) X^T \tilde{D} \tilde{D}^T X \\
&\quad + (1 + 4\alpha \cdot X^T X) \delta^T \tilde{D} \tilde{D}^T \delta + (1 + 4\alpha \cdot X^T X) d^T d \\
&\quad + 2 \cdot (1 - 4\alpha \cdot X^T X) \cdot r^T K_v^T \tilde{D}^T X \\
&\quad + 2 \cdot (1 + 4\alpha \cdot X^T X) \cdot r^T K_v^T \tilde{D}^T \delta + 2 \cdot (1 + 4\alpha \cdot X^T X) \cdot r^T K_v^T d \\
&\quad + 2 \cdot (1 - 4\alpha \cdot X^T X) X^T \tilde{D} \tilde{D}^T \delta + 2 \cdot (1 - 4\alpha \cdot X^T X) X^T \tilde{D} \cdot d \\
&\quad + 2 \cdot (1 + 4\alpha \cdot X^T X) \delta^T \tilde{D} \cdot d + 4 r^T K_v^T (D - \tilde{D})^T \Gamma^T X \| I - 4\alpha \cdot X X^T \| \\
&\quad + 4 \cdot X^T \tilde{D} \Gamma^T (D - \tilde{D})^T X \| I - 4\alpha \cdot X X^T \|
\end{aligned}
$$

$$+ 4\delta^T \tilde{D}(D - \tilde{D})^T \Gamma^T X \|I - 4\alpha \cdot XX^T\| + 4d^T(D - \tilde{D})^T \Gamma^T X \|I - 4\alpha \cdot XX^T\|$$

$$+ \frac{1}{\alpha} \text{tr}\{2 \cdot \tilde{D}^T \Gamma \|I - 4\alpha \cdot XX^T\| \hat{D} + \|I - 4\alpha \cdot XX^T\|^2 \hat{D}^T \Gamma^T \Gamma \hat{D}$$

$$- 8\alpha \cdot \tilde{D}^T XX^T \Gamma \|I - 4\alpha \cdot XX^T\| \hat{D}\}.$$

Bounding the trace term,

$$\frac{\|I - 4\alpha \cdot XX^T\|}{\alpha} \text{tr}\{2 \cdot \tilde{D}^T \Gamma \hat{D} + \|I - 4\alpha \cdot XX^T\| \hat{D}^T \Gamma^T \Gamma \hat{D} - 8\alpha \cdot \tilde{D}^T XX^T \Gamma \hat{D}\}$$

$$\leq -\frac{\|I - 4\alpha \cdot XX^T\|^2}{\alpha}\{\Gamma(2 - \Gamma)\|\tilde{D}\|^2 - 2 \cdot \Gamma(1 - \Gamma) D_{\max}\|\tilde{D}\| - \Gamma^2 D_{\max}^2\}.$$

Using (4.3.22), we have

$$(1 + 4\alpha \cdot X^T X)\delta^T \tilde{D}\tilde{D}^T \delta < 2 \cdot X^T \tilde{D}\tilde{D}^T X.$$

Then simplifying,

$$\begin{aligned}
\Delta J \leq &-r^T[I - (1 + 4\alpha \cdot X^T X)K_v^T K_v] \cdot r - (1 - 4\alpha \cdot X^T X)X^T \tilde{D}\tilde{D}^T X \\
&+ (1 + 4\alpha \cdot X^T X)d^T d + 2 \cdot (1 - 4\alpha \cdot X^T X) \cdot r^T K_v^T \tilde{D}^T X \\
&+ 2 \cdot (1 + 4\alpha \cdot X^T X) \cdot r^T K_v^T \tilde{D}^T \delta + 2 \cdot (1 + 4\alpha \cdot X^T X) \cdot r^T K_v^T d \\
&+ 2 \cdot (1 - 4\alpha \cdot X^T X)X^T \tilde{D}\tilde{D}^T \delta + 2 \cdot (1 - 4\alpha \cdot X^T X)X^T \tilde{D} \cdot d \\
&+ 2 \cdot (1 + 4\alpha \cdot X^T X)\delta^T \tilde{D} \cdot d - 4 \cdot \Gamma\|I - 4\alpha \cdot XX^T\|r^T K_v^T \tilde{D}^T X \\
&- 4 \cdot \Gamma\|I - 4\alpha \cdot XX^T\|\delta^T \tilde{D}\tilde{D}^T X - 4 \cdot \Gamma\|I - 4\alpha \cdot XX^T\|d^T \tilde{D}X \\
&+ 4 \cdot \Gamma K_{v\max}\|I - 4\alpha \cdot XX^T\| \cdot \|X\| \cdot D_{\max}\|r\| \\
&+ 4 \cdot \Gamma\|I - 4\alpha \cdot XX^T\| \cdot \|X\| \cdot \|\tilde{D}\|D_{\max}\|\delta\| \\
&+ 4 \cdot \Gamma\|I - 4\alpha \cdot X \cdot X^T\| \cdot \|X\| \cdot \|d\| \cdot D_{\max} \\
&+ 4 \cdot \Gamma X^T \tilde{D}D^T X\|I - 4\alpha \cdot XX^T\| \\
&- 4 \cdot \Gamma X^T \tilde{D}\tilde{D}^T X\|I - 4\alpha \cdot XX^T\| \\
&- \frac{\|I - 4\alpha \cdot XX^T\|^2}{\alpha}\{\Gamma(2 - \Gamma)\|\tilde{D}\|^2 - 2 \cdot \Gamma(1 - \Gamma)D_{\max}\|\tilde{D}\| - \Gamma^2 D_{\max}^2\}.
\end{aligned} \tag{4.A.6}$$

Completing squares for $\tilde{D}^T X$,

$$\begin{aligned}
\Delta J \leq &-(1 - \bar{\sigma}K_{v\max}^2)\|r\|^2 + 2 \cdot \gamma \cdot K_{v\max}\|r\| + \rho + 2 \cdot \bar{\sigma} \cdot K_{v\max}\|r\| \cdot \|X\| \cdot \|\tilde{D}\| \\
&- (1 - 4\alpha \cdot X^T X) \cdot \left\| \tilde{D}^T X - (K_v r + d + \tilde{D}^T \delta) \cdot \left(1 - \frac{2\Gamma\|I - 4\alpha \cdot XX^T\|}{1 - 4\alpha \cdot X^T X}\right) \right\|^2 \\
&- \rho_1\|\tilde{D}\|^2 + 2 \cdot \rho_2\|\tilde{D}\|,
\end{aligned} \tag{4.A.7}$$

where

$$\begin{aligned}
\bar{\sigma} &= 1 + 4\alpha \cdot X^T X + \frac{[2 \cdot \Gamma\|I - 4\alpha \cdot XX^T\| - (1 - 4\alpha \cdot X^T X)]^2}{1 - 4\alpha \cdot X^T X} \\
&= 2 - 8\alpha \cdot X^T X - 4 \cdot \Gamma\|I - 4\alpha \cdot XX^T\| + \frac{4 \cdot \Gamma^2\|I - 4\alpha \cdot XX^T\|^2}{1 - 4\alpha \cdot X^T X},
\end{aligned} \tag{4.A.8}$$

$$\gamma = \bar{\sigma}\|d\| + 2 \cdot \Gamma(1 - 4\alpha \cdot X_{\max}^2) \cdot X_{\max} D_{\max}, \tag{4.A.9}$$

$$\rho = \bar{\sigma}\|d\|^2 + 4 \cdot \Gamma(1 - 4\alpha \cdot X_{\max}^2) \cdot X_{\max}\|d\| D_{\max} + \frac{\|I - 4\alpha \cdot XX^T\|^2}{\alpha}\Gamma^2 D_{\max}^2, \tag{4.A.10}$$

$$\rho_1 = \frac{\|I - 4\alpha \cdot XX^T\|^2}{\alpha}\Gamma(2 - \Gamma) + 4 \cdot \Gamma\|I - 4\alpha \cdot XX^T\| \cdot \|X\|^2$$
$$- \frac{(1 - 4\alpha \cdot X^TX - 2\Gamma\|I - 4\alpha \cdot XX^T\|)^2}{1 - 4\alpha \cdot X^TX}\|X\|^2, \tag{4.A.11}$$

$$\rho_2 = 4 \cdot \Gamma\|I - 4\alpha \cdot XX^T\| \cdot \|X\|^2 D_{\max} + \frac{\|I - 4\alpha \cdot XX^T\|^2}{\alpha}\Gamma(1 - \Gamma)D_{\max} + \bar{\sigma}\|d\| \cdot \|X\|. \tag{4.A.12}$$

Simplifying,

$$\Delta J \leq -(1 - \bar{\sigma} K_{v\,\max}^2)\|r\|^2 + 2 \cdot \gamma \cdot K_{v\,\max}\|r\| + \rho$$
$$+ 2\bar{\sigma} K_{v\,\max}\|r\| \cdot \|X\| \cdot \|\tilde{D}\| - \rho_1\|\tilde{D}\|^2 + 2 \cdot \rho_2\|\tilde{D}\|. \tag{4.A.13}$$

Completing squares for $\|\tilde{D}\|$ in (4.A.13),

$$\Delta J \leq -(1 - \bar{\sigma} K_{v\,\max}^2)\|r\|^2 + 2 \cdot \gamma \cdot K_{v\,\max}\|r\|$$
$$+ \rho - \rho_1\left[\|\tilde{D}\| - \frac{\rho_2 + \bar{\sigma} K_{v\,\max}\|r\| \cdot \|X\|}{\rho_1}\right]^2 + \frac{[\rho_2 + \bar{\sigma} K_{v\,\max}\|r\| \cdot \|X\|]^2}{\rho_1},$$

$$\Delta J \leq -(1 - \bar{\sigma} K_{v\,\max}^2)\|r\|^2 + 2 \cdot \gamma \cdot K_{v\,\max}\|r\|$$
$$+ \rho - \rho_1\left[\|\tilde{D}\| - \frac{\rho_2 + \bar{\sigma} K_{v\,\max}\|r\| \cdot \|X\|}{\rho_1}\right]^2$$
$$+ \frac{\rho_2^2}{\rho_1} + \frac{2\rho_2\bar{\sigma} K_{v\,\max}\|r\| \cdot X_{\max}}{\rho_1} + \frac{\bar{\sigma}^2 K_{v\,\max}^2\|r\|^2 \cdot X_{\max}^2}{\rho_1},$$

$$\Delta J \leq -\left[1 - \left(1 + \frac{\bar{\sigma} \cdot X_{\max}^2}{\rho_1}\right)\bar{\sigma} \cdot K_{v\,\max}^2\right]\|r\|^2 + 2 \cdot K_{v\,\max}\|r\| \cdot \left(\gamma + \frac{\rho_2\bar{\sigma} X_{\max}}{\rho_1}\right) + \rho$$
$$- \rho_1\left[\|\tilde{D}\| - \frac{\rho_2 + \bar{\sigma} K_{v\,\max}\|r\| \cdot \|X\|}{\rho_1}\right]^2 + \frac{\rho_2^2}{\rho_1},$$

$$\Delta J \leq -\left[1 - \left(1 + \frac{\bar{\sigma} \cdot X_{\max}^2}{\rho_1}\right)\bar{\sigma} \cdot K_{v\,\max}^2\right]\|r\|^2$$
$$+ 2 \cdot K_{v\,\max}\|r\| \cdot \left(\gamma + \frac{\rho_2\bar{\sigma} X_{\max}}{\rho_1}\right) + \rho + \frac{\rho_2^2}{\rho_1}. \tag{4.A.14}$$

Completing squares for $\|r\|$ in (4.A.14),

$$\Delta J \leq -\left[1 - \left(1 + \frac{\bar{\sigma} \cdot X_{\max}^2}{\rho_1}\right)\bar{\sigma} \cdot K_{v\,\max}^2\right]\left[\|r\| - \frac{K_{v\,\max}\left(\gamma + \frac{\rho_2\bar{\sigma} \cdot X_{\max}}{\rho_1}\right)}{1 - \left(1 + \frac{\bar{\sigma} \cdot X_{\max}^2}{\rho_1}\right)\bar{\sigma} \cdot K_{v\,\max}^2}\right]^2$$

$$+ \rho + \frac{K_{v\,\max}^2\left(\gamma + \frac{\rho_2\bar{\sigma} \cdot X_{\max}}{\rho_1}\right)^2}{1 - \left(1 + \frac{\bar{\sigma} \cdot X_{\max}^2}{\rho_1}\right)\bar{\sigma} \cdot K_{v\,\max}^2} + \frac{\rho_2^2}{\rho_1}.$$

Then $\Delta J < 0$ if

$$\left[\|r\| - \frac{K_{v\max}\left(\gamma + \frac{\rho_2\bar{\sigma}\cdot X_{\max}}{\rho_1}\right)}{1 - \left(1 + \frac{\bar{\sigma}\cdot X_{\max}^2}{\rho_1}\right)\bar{\sigma}\cdot K_{v\max}^2}\right]^2$$

$$\geq \frac{K_{v\max}^2\left(\gamma + \frac{\rho_2\bar{\sigma}\cdot X_{\max}}{\rho_1}\right)^2 + \left[\rho + \frac{\rho_2^2}{\rho_1}\right]\left[1 - \left(1 + \frac{\bar{\sigma}\cdot X_{\max}^2}{\rho_1}\right)\bar{\sigma}\cdot K_{v\max}^2\right]}{\left[1 - \left(1 + \frac{\bar{\sigma}\cdot X_{\max}^2}{\rho_1}\right)\bar{\sigma}\cdot K_{v\max}^2\right]^2},$$

$$\|r\| \geq \frac{K_{v\max}\left(\gamma + \frac{\rho_2\bar{\sigma}\cdot X_{\max}}{\rho_1}\right)}{1 - \left(1 + \frac{\bar{\sigma}\cdot X_{\max}^2}{\rho_1}\right)\bar{\sigma}\cdot K_{v\max}^2}$$

$$+ \frac{\sqrt{K_{v\max}^2\left(\gamma + \frac{\rho_2\bar{\sigma}\cdot X_{\max}}{\rho_1}\right)^2 + \left[\rho + \frac{\rho_2^2}{\rho_1}\right]\left[1 - \left(1 + \frac{\bar{\sigma}\cdot X_{\max}^2}{\rho_1}\right)\bar{\sigma}\cdot K_{v\max}^2\right]}}{1 - \left(1 + \frac{\bar{\sigma}\cdot X_{\max}^2}{\rho_1}\right)\bar{\sigma}\cdot K_{v\max}^2},$$

$$\|r\| \geq \tag{4.A.15}$$

$$\frac{K_{v\max}(\gamma\rho_1 + \rho_2\bar{\sigma}X_{\max}) + \sqrt{K_{v\max}^2(\gamma\rho_1 + \rho_2\bar{\sigma}X_{\max})^2 + (\rho\rho_1 + \rho_2^2)(\rho_1 - (\rho_1 + \bar{\sigma}X_{\max}^2)\bar{\sigma}K_{v\max}^2)}}{\rho_1 - (\rho_1 + \bar{\sigma}X_{\max}^2)\bar{\sigma}K_{v\max}^2}.$$

Similarly, completing squares for $\|r\|$ in (4.A.13) yields

$$\Delta J \leq -(1 - \bar{\sigma}K_{v\max}^2)\left[\|r\| - \frac{\gamma\cdot K_{v\max} + \bar{\sigma}K_{v\max}X_{\max}\cdot\|\tilde{D}\|}{1 - \bar{\sigma}K_{v\max}^2}\right]^2$$

$$+ \rho + \frac{(\gamma\cdot K_{v\max} + \bar{\sigma}K_{v\max}X_{\max}\cdot\|\tilde{D}\|)^2}{1 - \bar{\sigma}K_{v\max}^2} - \rho_1\|\tilde{D}\|^2 + 2\cdot\rho_2\|\tilde{D}\|,$$

$$\Delta J \leq \rho + \frac{\gamma^2\cdot K_{v\max}^2}{1 - \bar{\sigma}K_{v\max}^2} - \left(\rho_1 - \frac{\bar{\sigma}^2 K_{v\max}^2 X_{\max}^2}{1 - \bar{\sigma}K_{v\max}^2}\right)\|\tilde{D}\|^2$$

$$+ 2\cdot\left(\rho_2 + \frac{\gamma\cdot K_{v\max}^2\bar{\sigma}\cdot X_{\max}}{1 - \bar{\sigma}K_{v\max}^2}\right)\|\tilde{D}\|. \tag{4.A.16}$$

Completing squares for $\|\tilde{D}\|$ in (4.A.16),

$$\Delta J \leq \rho + \frac{\gamma^2\cdot K_{v\max}^2}{1 - \bar{\sigma}K_{v\max}^2} + \frac{\left(\rho_2 + \frac{\gamma\cdot K_{v\max}^2\bar{\sigma}\cdot X_{\max}}{1 - \bar{\sigma}K_{v\max}^2}\right)^2}{\rho_1 - \frac{\bar{\sigma}^2 K_{v\max}^2 X_{\max}^2}{1 - \bar{\sigma}K_{v\max}^2}}$$

$$- \left(\rho_1 - \frac{\bar{\sigma}^2 K_{v\max}^2 X_{\max}^2}{1 - \bar{\sigma}K_{v\max}^2}\right)\left[\|\tilde{D}\| - \frac{\rho_2 + \frac{\gamma\cdot K_{v\max}^2\bar{\sigma}\cdot X_{\max}}{1 - \bar{\sigma}K_{v\max}^2}}{\rho_1 - \frac{\bar{\sigma}^2 K_{v\max}^2 X_{\max}^2}{1 - \bar{\sigma}K_{v\max}^2}}\right]^2,$$

Then $\Delta J < 0$ if

$$\left(\rho_1 - \frac{\bar{\sigma}^2 K_{v\max}^2 X_{\max}^2}{1 - \bar{\sigma}K_{v\max}^2}\right)\left[\|\tilde{D}\| - \frac{\rho_2 + \frac{\gamma\cdot K_{v\max}^2\bar{\sigma} X_{\max}}{1 - \bar{\sigma}K_{v\max}^2}}{\rho_1 - \frac{\bar{\sigma}^2 K_{v\max}^2 X_{\max}^2}{1 - \bar{\sigma}K_{v\max}^2}}\right]^2$$

$$> \rho + \frac{\gamma^2 K_{v\max}^2}{1 - \bar{\sigma}K_{v\max}^2} + \frac{\left(\rho_2 + \frac{\gamma\cdot K_{v\max}^2\bar{\sigma} X_{\max}}{1 - \bar{\sigma}K_{v\max}^2}\right)^2}{\rho_1 - \frac{\bar{\sigma}^2 K_{v\max}^2 X_{\max}^2}{1 - \bar{\sigma}K_{v\max}^2}},$$

$$\|\tilde{D}\| > \frac{\rho_2 + \frac{\gamma \cdot K_{v\,\max}^2 \bar{\sigma} \cdot X_{\max}}{1 - \bar{\sigma} K_{v\,\max}^2}}{\rho_1 - \frac{\bar{\sigma}^2 K_{v\,\max}^2 X_{\max}^2}{1 - \bar{\sigma} K_{v\,\max}^2}} + \sqrt{\frac{\rho + \frac{\gamma^2 \cdot K_{v\,\max}^2}{1 - \bar{\sigma} K_{v\,\max}^2}}{\left(\rho_1 - \frac{\bar{\sigma}^2 K_{v\,\max}^2 X_{\max}^2}{1 - \bar{\sigma} K_{v\,\max}^2}\right)} + \frac{\left(\rho_2 + \frac{\gamma \cdot K_{v\,\max}^2 \bar{\sigma} \cdot X_{\max}}{1 - \bar{\sigma} K_{v\,\max}^2}\right)^2}{\left(\rho_1 - \frac{\bar{\sigma}^2 K_{v\,\max}^2 X_{\max}^2}{1 - \bar{\sigma} K_{v\,\max}^2}\right)^2}},$$

$$\|\tilde{D}\| > \frac{\rho_2(1 - \bar{\sigma} K_{v\,\max}^2) + \gamma \cdot K_{v\,\max}^2 \bar{\sigma} \cdot X_{\max}}{\rho_1(1 - \bar{\sigma} K_{v\,\max}^2) - \bar{\sigma}^2 K_{v\,\max}^2 X_{\max}^2} \tag{4.A.17}$$

$$+ \sqrt{\frac{\rho(1 - \bar{\sigma} K_{v\,\max}^2) + \gamma^2 \cdot K_{v\,\max}^2}{\rho_1(1 - \bar{\sigma} K_{v\,\max}^2) - \bar{\sigma}^2 K_{v\,\max}^2 X_{\max}^2} + \frac{(\rho_2(1 - \bar{\sigma} K_{v\,\max}^2) + \gamma \cdot K_{v\,\max}^2 \bar{\sigma} \cdot X_{\max})^2}{(\rho_1(1 - \bar{\sigma} K_{v\,\max}^2) - \bar{\sigma}^2 K_{v\,\max}^2 X_{\max}^2)^2}}.$$

From (4.A.15) or (4.A.17), ΔJ is negative outside a compact set. According to a standard Lyapunov theorem extension, it can be concluded that the tracking error $r(k)$ and the error in deadzone estimates $\tilde{D}(k)$ are GUUB. $\square$

Chapter 5

Neural Control of Systems with Backlash

In this chapter, we show how to use NN learning systems to compensate for the effects of backlash in industrial motion systems. We discuss both continuous-time and discrete-time NN compensators. Discrete-time compensators, of course, are required for digital control purposes on actual industrial systems. The NN backlash compensators must appear in the feedforward control loop, not the feedback loop, which presents difficulties in designing the controller structure and learning laws. Since backlash is a dynamic effect, one requires a dynamic compensator. The NN compensator is based on the dynamic inversion approach, which dynamically inverts the backlash to nullify its effect. Rigorous proofs of closed-loop stability for the backlash compensators are provided and yield tuning algorithms for the weights of the NNs.

5.1   Introduction to backlash control

The backlash nonlinearity was discussed in Chapter 2. It is depicted in Figure 5.1.1. Backlash is a nonlinearity in mechanical systems caused by the imperfection of the gears or some other mechanical parts. It results in a delay in the system motion. Backlash can reduce the system stability and generate noise and undesired vibrations. The uncertainty caused by the backlash can also decrease the repeatability and accuracy of mechanical systems. Compared with deadzone, which is a static nonlinearity, backlash is a dynamic nonlinearity (i.e., it has its own internal dynamics). Common industrial control systems often have the structure of a backlash nonlinearity followed by a nonlinear dynamical system. Examples include xy-positioning tables (Li and Cheng (1996)), robot manipulators (Lewis, Abdallah, and Dawson (1993)), and overhead crane mechanisms. Backlash is particularly common in actuators, such as mechanical connections, hydraulic servovalves, and electric servomotors. There are many applications, such as instrument differential gear trains and servomechanisms, that require the complete elimination of backlash in order to function properly.

Due to the nonanalytic nature of the actuator nonlinearities and the fact that their exact nonlinear functions are unknown, such systems present a challenge for the control design engineer. Many mechanical solutions have been developed to overcome backlash, for example, spring-loaded split-gear assemblies and dual motor systems. These mechanical solutions can satisfactorily handle the backlash problem, but they give rise to other problems like decreased accuracy and reduced bandwidth. They are also expensive and energy consuming and increase the overall weight of the system. A backlash compensation scheme not based on mechanical devices would be more convenient.

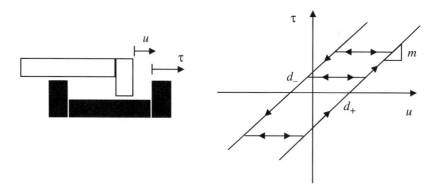

Figure 5.1.1. *Backlash nonlinearity.*

Standard control techniques like PD or proportional-integral-derivative (PID) are not very effective for backlash compensation since, when backlash is present, they can produce limit cycles. Furthermore, closed-loop stability is difficult to prove when backlash is present. Backlash compensation has been studied by many researchers. Rigorous results for motion tracking of such systems are notably sparse, though ad hoc techniques relying on simulations for verification of effectiveness are prolific. A method of determining the dynamic effects caused by backlash in mechanisms was given in Goodman (1963). A model of the backlash for spur-gear systems was given in Shing, Tsai, and Krishnaprasad (1993). Compensation schemes based on adaptive control were given in Tao and Kokotović (1995, 1996). NN compensation of a gear backlash hysteresis was given in Seidl et al. (1995).

The use of NNs in feedback control applications has accelerated in recent years. Particularly important in NN control are the *universal function approximation capabilities* of NN systems (see Chapter 1). NN systems offer significant advantages over adaptive control, including no requirement for linearity in the parameter assumptions and no need to compute a regression matrix for each specific system. Some references for NN in feedback control are discussed at the beginning of Chapter 4 and appear in the bibliography. Many of these contain rigorous proofs of stability and performance.

We propose in this chapter to use NN for the compensation of backlash in industrial systems. Compensation of a dynamic nonlinearity such as backlash requires the design of a dynamic compensator. We propose designing a compensator with dynamics using the *dynamic inversion* of the system nonlinearity. Inverse dynamics control is closely related to the so-called backstepping technique (Kanellakopoulos, Kokotović, and Morse (1991)). An inverse dynamics approach using adaptive and robust control techniques was presented in Song, Mitchell, and Lai (1994). Dynamic inversion for flight control design was given in Lane and Stengel (1988), Enns et al. (1994), and Stevens and Lewis (2001). Modeling inverse dynamics in the feedforward path using recurrent NNs was presented in Yan and Li (1997). Seminal work on dynamic inversion using NN was presented in Kim and Calise (1997), Leitner, Calise, and Prasad (1997), and McFarland and Calise (2000), where an NN was used for cancellation of the inversion error. We plan to use an approach based on this work in this chapter.

5.2 Continuous-time neural network backlash compensation

A general model of the backlash is assumed here, and it is not required to be symmetric. We use a specialized backstepping approach known as dynamics inversion to derive a compensator that employs an NN in the feedforward loop. This is unusual in NN control since the NN usually appears in a feedback loop. The proposed method can be applied for compensation of a large class of right-invertible dynamical nonlinearities. The generality of the method and its applicability to a broad range of nonlinear functions make this approach a useful tool for compensation of backlash, hysteresis, etc.

5.2.1 Backlash inverse

The backlash nonlinearity is shown in Figure 5.1.1, and a mathematical model is given (Tao and Kokotović (1996)) by

$$\dot{\tau} = B(\tau, u, \dot{u}) = \begin{cases} m\dot{u} & \text{if } \dot{u} > 0 \text{ and } \tau = mu - md_+, \\ & \text{if } \dot{u} < 0 \text{ and } \tau = mu - md_-, \\ 0 & \text{otherwise.} \end{cases} \qquad (5.2.1)$$

One can see that backlash is a first-order velocity-driven dynamical system with inputs u and $\dot{u}$ and state τ. Since the backlash dynamics is given in terms of the derivative of the output $\tau(t)$, it contains its own dynamics; therefore, its compensation requires the design of a dynamic compensator.

Whenever the motion $u(t)$ changes its direction, the motion $\tau(t)$ is delayed from the motion of $u(t)$. The objective of a backlash compensator is to make this delay as small as possible (i.e., to make the $\tau(t)$ closely follow $u(t)$). In order to cancel the effect of backlash in the system, the backlash precompensator needs to generate the inverse of the backlash nonlinearity. The backlash inverse function is shown in Figure 5.2.1. Thus if the signal $w(t)$ enters the backlash inverse and generates signal $u(t)$, which is subsequently sent into the backlash to produce $\tau(t)$, the throughput in the ideal case from $w(t)$ to $\tau(t)$ would be equal to unity. This assumes that the backlash parameters d_+, d_-, and m used in the backlash inverse are exactly correct.

The dynamics of the proposed backlash compensator are given by

$$\dot{u} = B_{\text{inv}}(u, w, \dot{w}). \qquad (5.2.2)$$

The backlash inverse characteristic shown in Figure 5.2.1 can be decomposed into two functions: a direct feedforward term plus the additional *modified backlash inverse* term as

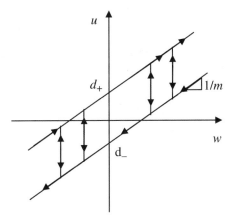

Figure 5.2.1. *Backlash inverse.*

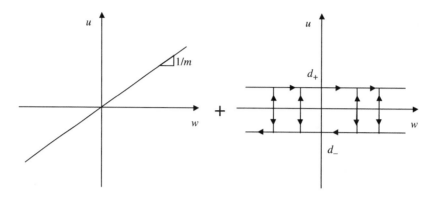

Figure 5.2.2. *Backlash inverse decomposition with unity feedforward path.*

shown in Figure 5.2.2. This decomposition allows design of a compensator that has a better
structure when used in the feedforward path. In this chapter, we plan to use an NN in the
backlash inverse block to adaptively estimate the backlash parameters by online learning.

5.2.2 Dynamics and control of nonlinear motion systems

The dynamics of a large class of single-input nonlinear industrial systems can be written in
the Brunovsky form

$$\dot{x}_1 = x_2,$$
$$\dot{x}_2 = x_3,$$
$$\vdots \tag{5.2.3}$$
$$\dot{x}_n = f(x) + \tau_d + \tau,$$
$$y = x_1,$$

where the output is $y(t)$, the state is $x \equiv [x_1 \quad x_2 \quad \cdots \quad x_n]^T$, τ_d is a disturbance, $\tau(t)$ is
the actuator output, and the function $f(x)$ represents system nonlinearities like friction. The
actuator output $\tau(t)$ is related to the control input $u(t)$ through the backlash nonlinearity
(5.2.1). Therefore, the overall dynamics of the system consists of (5.2.3) and backlash
dynamics (5.2.1).

The following assumptions are needed. They are typical assumptions commonly made
in the literature and they hold for most practical systems.

Assumption 1 (Bounded disturbance). The unknown disturbance satisfies $\|\tau_d\| \leq \tau_M$,
with $\tau_M(t)$ a known positive constant.

Assumption 2 (Bounded estimation error). The nonlinear function $f(x)$ is assumed to be
unknown, but a fixed estimate $\hat{f}(x)$ is assumed known such that the functional estimation
error $\tilde{f}(x) = f(x) - \hat{f}(x)$ satisfies

$$\|\tilde{f}(x)\| \leq f_M(x) \tag{5.2.4}$$

for some known bounding function $f_M(x)$.

Assumption 2 is standard in robust control (Corless and Leitmann (1982)). Note that in practical systems the bound $f_M(x)$ can be computed knowing the upper bound on payload masses, frictional effects, and so on. One could estimate the function $f(x)$ using adaptive or NN techniques (Lewis, Yesildirek, and Liu (1996)), but adding that functionality in this chapter would distract from our major topic of concern, which is NN compensation of backlash.

To design a motion controller that causes the system output $y(x)$ to track a smooth prescribed trajectory $y_d(x)$, we define the desired state trajectory as

$$x_d(t) = [y_d \quad \dot{y}_d \quad \cdots \quad y_d^{(n-1)}]^T, \tag{5.2.5}$$

with $y_d^{(n-1)}$ the $(n-1)$st derivative. We define the *tracking error* by

$$e = x - x_d \tag{5.2.6}$$

and the *filtered tracking error* by

$$r = [\lambda_1 \quad \lambda_2 \quad \cdots \quad \lambda_{n-1} \quad 1]e \equiv [\Lambda^T \quad 1]e, \tag{5.2.7}$$

with Λ a gain parameter vector selected so that $e(t) \to 0$ exponentially as $r(t) \to 0$. Then (5.2.7) is a stable system so that $e(t)$ is bounded as long as the controller guarantees that the filtered error $r(t)$ is bounded.

Differentiating (5.2.7) and invoking (5.2.3), it can be seen that the dynamics are expressed in terms of the filtered error as

$$\dot{r} = f(x) + Y_d + \tau_d + \tau, \tag{5.2.8}$$

where

$$Y_d = -y_d^{(n)} + [0 \quad \Lambda^T]e \tag{5.2.9}$$

is a known function of the desired trajectory and actual states.

The next assumption always holds.

Assumption 3 (Bounded desired trajectory). The desired trajectory is bounded so that

$$\|x_d(t)\| \le X_d, \tag{5.2.10}$$

where X_d is a known constant.

Ideal control law assuming no backlash

Let us first design an ideal control law supposing that there is no backlash. In the next subsection, we will convert this to a realistic controller by adding backlash compensation. A robust compensation scheme for unknown terms in $f(x)$ is provided by selecting the tracking controller

$$\tau_{\text{des}} = -K_v r - \hat{f}(x) - Y_d + v_1, \tag{5.2.11}$$

with $\hat{f}(x)$ an estimate for the nonlinear terms $f(x)$ and $v_1(t)$ a robustifying term to be selected for disturbance rejection. The feedback gain matrix $K_v > 0$ is often selected diagonal. The estimate $\hat{f}(x)$ is fixed in this paper and will not be adapted, as is common in robust control (Corless and Leitmann (1982)).

The next theorem is the first step in the backstepping design; it shows that if there is no backlash, the desired control law (5.2.11) will keep the filtered tracking error small in the sense of uniform ultimate boundedness (see Chapter 2).

Theorem 5.2.1 (Control law for outer tracking loop). *Given the system* (5.2.3) *and Assumptions 1 and 2, select the tracking control law* (5.2.11). *Choose the robustifying signal v_1 as*

$$v_1(t) = -(f_M(x) + \tau_M)\frac{r}{\|r\|}. \tag{5.2.12}$$

Then the filtered tracking error $r(t)$ is UUB, and it can be kept as small as desired by increasing the gains K_v.

Proof. Select the Lyapunov function candidate

$$L_1 = \frac{1}{2}r^T r.$$

Differentiating L_1 and using (5.2.8) yields

$$\dot{L}_1 = r^T(f(x) + Y_d + \tau_d + \tau_{\text{des}}).$$

Applying the tracking control law (5.2.11), one has

$$\dot{L}_1 = r^T(f(x) + Y_d + \tau_d - K_V r - \hat{f}(x) - Y_d + v_1),$$
$$\dot{L}_1 = r^T(\tilde{f}(x) + \tau_d - K_V r + v_1),$$
$$\dot{L}_1 = -r^T K_V r + r^T\left(\tilde{f}(x) + \tau_d - (f_M(x) + \tau_M)\frac{r}{\|r\|}\right).$$

This expression may be bounded as

$$\dot{L}_1 \leq -K_{V\min}\|r\|^2 - \|r\|(f_M + \tau_M) + \|r\|\|\tilde{f} + \tau_d\|.$$

Using Assumptions 1 and 2, one can conclude that $\dot{L}_1$ is guaranteed negative as long as $\|r\| \neq 0$. □

5.2.3 Neural network backlash compensation using dynamic inversion

We have just shown that the ideal robust control law (5.2.11) will stabilize the system and yield good tracking if there is no backlash. Now we must show how to modify this control law if there is backlash.

An NN dynamic backlash compensator is now designed using the backstepping technique originally developed by Kanellakopoulos, Kokotović, and Morse (1991). A specific form of backstepping known as dynamic inversion (Enns et al. (1994)) is used here in order to ensure that the compensator has dynamics. We shall show how to tune or learn the weights of the NN online so that the tracking error is guaranteed small and all internal states (e.g., the NN weights) are bounded. It is assumed that the actuator output $\tau(t)$ is measurable.

Backstepping for backlash compensation has two steps. The first step is to design an ideal control law assuming no backlash, which we have just accomplished. Unfortunately, in the presence of the unknown backlash nonlinearity, the desired value (5.2.11) and the actual value of the control signal $\tau(t)$ sent to the process will be different, since $\tau(t)$ is the output of the backlash. The goal of the second step in backstepping controller design is to ensure that the actual actuator output follows the desired actuator output even in the presence of the backlash nonlinearity, thus achieving backlash compensation. Following the idea of dynamic inversion using an NN for compensation of the inversion error as originally given by Calise et al. (Kim and Calise (1997), McFarland and Calise (2000)), we design an NN backlash compensator and give a rigorous analysis of the closed-loop system stability.

To complete the second step in backstepping design, proceed as follows. The actuator output given by (5.2.11) is the desired, ideal signal. In order to find the complete system error dynamics including the backlash, define the error between the desired and actual actuator outputs as

$$\tilde{\tau} = \tau_{\text{des}} - \tau. \tag{5.2.13}$$

Differentiating, one has

$$\begin{aligned}
\dot{\tilde{\tau}} &= \dot{\tau}_{\text{des}} - \dot{\tau} \\
&= \dot{\tau}_{\text{des}} - B(\tau, u, \dot{u}),
\end{aligned} \tag{5.2.14}$$

which together with (5.2.8) and involving (5.2.11) represents the complete system error dynamics.

The dynamics of the backlash nonlinearity can be written as

$$\dot{\tau} = \varphi, \tag{5.2.15}$$

$$\varphi = B(\tau, u, \dot{u}), \tag{5.2.16}$$

where $\varphi(t)$ is the pseudocontrol input (Leitner et al. (1997)). In the case where the backlash parameters are known, the ideal backlash inverse is given by

$$\dot{u} = B^{-1}(u, \tau, \varphi). \tag{5.2.17}$$

However, since the backlash and therefore its inverse are not known, one can only approximate the backlash inverse as

$$\dot{\hat{u}} = \hat{B}^{-1}(\hat{u}, \tau, \hat{\varphi}). \tag{5.2.18}$$

The backlash dynamics can now be written as

$$\begin{aligned}
\dot{\tau} &= B(\tau, \hat{u}, \dot{\hat{u}}) \\
&= \hat{B}(\tau, \hat{u}, \dot{\hat{u}}) + \tilde{B}(\tau, \hat{u}, \dot{\hat{u}}) \\
&= \hat{\varphi} + \tilde{B}(\tau, \hat{u}, \dot{\hat{u}}),
\end{aligned} \tag{5.2.19}$$

where

$$\hat{\varphi} = \hat{B}(\tau, \hat{u}, \dot{\hat{u}}), \tag{5.2.20}$$

and therefore its inverse is

$$\dot{\hat{u}} = \hat{B}^{-1}(\tau, \hat{u}, \hat{\varphi}). \qquad (5.2.21)$$

The unknown function $\tilde{B}(\tau, \hat{u}, \dot{\hat{u}})$, which represents the backlash inversion error, will be approximated using an NN. Based on the NN approximation property, the backlash inversion error can be represented by an NN as

$$\tilde{B}(\tau, \hat{u}, \dot{\hat{u}}) = W^T \sigma(V^T x_{\text{NN}}) + \varepsilon(x_{\text{NN}}), \qquad (5.2.22)$$

where $\varepsilon(x_{\text{NN}})$ is the NN approximation error bounded on a compact set. Note that $\tilde{B}(\tau, \hat{u}, \dot{\hat{u}})$ is not a smooth function since the backlash inverse itself is not smooth. One should therefore use the augmented NN described in Chapter 3, which can approximate functions with jumps very well. The ideal NN weights that give a good approximation are denoted by W and V. They are assumed to be unknown.

The output y_{NN} of the NN is given by

$$y_{\text{NN}} = \hat{W}^T \sigma(\hat{V}^T x_{\text{NN}}), \qquad (5.2.23)$$

where $\hat{W}$ and $\hat{V}$ are the actual NN weights in the circuit. They are given by the tuning algorithms to be determined. Presumably, for good performance, they should be close in some sense to the ideal weights W and V.

The NN input vector is chosen as $x_{\text{NN}} = [1 \quad r^T \quad x_d^T \quad \tilde{\tau}^T \quad \tau \quad y_{\text{NN}} \quad \|\hat{Z}\|_F^2]^T$, which is chosen based on the functional dependence of the function $\tilde{B}(\tau, \hat{u}, \dot{\hat{u}})$. Note that the network output is also the NN input. This requires a fixed-point solution assumption, which holds for bounded sigmoidal activation functions (Kim and Calise (1997), McFarland and Calise (2000)).

The matrices $\hat{W}$ and $\hat{V}$ are estimates of the ideal NN weights, which are given by the NN tuning algorithms. Define the weight estimation errors as

$$\tilde{V} = V - \hat{V}, \qquad \tilde{W} = W - \hat{W}, \qquad \tilde{Z} = Z - \hat{Z}, \qquad (5.2.24)$$

and the hidden-layer output error for a given x as

$$\tilde{\sigma} = \sigma - \hat{\sigma} \equiv \sigma(V^T x_{\text{NN}}) - \sigma(\hat{V}^T x_{\text{NN}}). \qquad (5.2.25)$$

In order to design a stable closed-loop system with backlash compensation, one selects the nominal backlash inverse $\dot{\hat{u}} = \hat{B}^{-1}(\hat{u}, \tau, \hat{\varphi}) = \hat{\varphi}$ and pseudocontrol input $\hat{\varphi}$ as

$$\hat{\varphi} = K_b \tilde{\tau} + \dot{\tau}_{\text{des}} - \hat{W}^T \sigma(\hat{V}^T x_{\text{NN}}) + v_2, \qquad (5.2.26)$$

where $v_2(t)$ is a robustifying term detailed later.

Note that we do not require two NNs as in the deadzone compensator in Chapter 4, since we assume that the signal τ is measurable. If τ is not measurable, then the compensator structure would involve two NNs, one as a backlash estimator and the other as a compensator.

Figure 5.2.3 shows the closed-loop system with the NN backlash compensator. The proposed backlash compensation scheme consists of a feedforward-path, direct feed term (with gain K_b) plus the error term estimated by NN. This is a result of the decomposition performed in Figure 5.2.2. Note that this means there are three control loops: one PD control loop with gains K_v and K_b that does not involve the NN, a parallel NN feedforward loop,

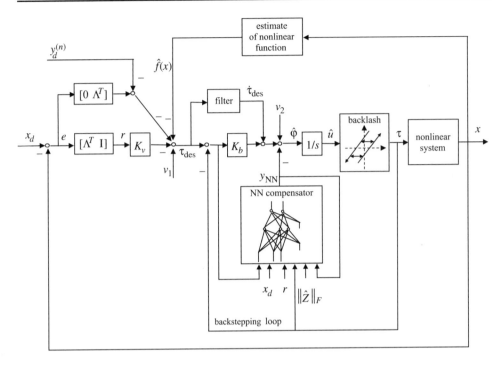

Figure 5.2.3. *NN backlash compensator.*

and an inner feedback linearization loop that provides the estimate $\hat{f}(x)$. It is vital to have the PD loop since then it is very easy to initialize the NN weights, as will be discussed.

Using the proposed controller (5.2.26), the error dynamics (5.2.14) can be written as

$$\dot{\tilde{\tau}} = \dot{\tau}_{\text{des}} - \hat{\varphi} - \tilde{B}(\tau, \hat{u}, \dot{\hat{u}})$$
$$= -K_b\tilde{\tau} + \hat{W}^T\sigma(\hat{V}^T x_{\text{NN}}) - v_2 - W^T\sigma(V^T x_{\text{NN}}) - \varepsilon. \tag{5.2.27}$$

Overcoming the linearity-in-the-parameters restriction

Note that the first-layer weights V appear in nonlinear fashion in (5.2.27). This makes control design exceedingly difficult. Fortunately, one can use the Taylor series expansion in order to handle this problem. This means that we are able to overcome the standard restriction of linearity in the tunable parameters that is generally required in adaptive control approaches. This is a key factor that makes our NN approach more powerful than adaptive control for backlash compensation.

Applying the Taylor series as developed in Lewis, Yesildirek, and Liu (1996), one obtains the error dynamics

$$\dot{\tilde{\tau}} = -K_b\tilde{\tau} - \tilde{W}^T(\hat{\sigma} - \hat{\sigma}'\hat{V}^T x_{\text{NN}}) - \hat{W}^T\hat{\sigma}'\tilde{V}^T x_{\text{NN}} + w - v_2, \tag{5.2.28}$$

where the disturbance term is given by

$$w = -\tilde{W}^T\hat{\sigma}'V^T x_{nn} - W^T O(\tilde{V}^T x_{\text{NN}})^2 - \varepsilon(x_{\text{NN}}), \tag{5.2.29}$$

with $O(\tilde{V}^T x_{\text{NN}})^2$ denoting the higher order terms in Taylor series expansion. Assuming that the approximation property of the NN holds, the norm of the disturbance term can be bounded above as (Lewis, Yesildirek, and Liu (1996))

$$\|w\| \leq V_M \|\tilde{W}\|_F \|x_{\text{NN}}\| + c_1 + c_2 \|\tilde{V}\|_F \|x_{\text{NN}}\| + \varepsilon_N, \tag{5.2.30}$$

where c_1 and c_2 are positive computable constants. The NN input is bounded by

$$\|x_{\text{NN}}\| \leq c_3 + \|r\| + X_d + \|\tilde{\tau}\| + c_4 \|\tilde{Z}\|_F. \tag{5.2.31}$$

Combining these two inequalities, one has the disturbance term bounded by

$$\|w\| \leq (V_M \|\tilde{W}\|_F + c_2 \|\tilde{V}\|_F)(c_3 + \|r\| + X_d + \|\tilde{\tau}\| + c_4 \|\tilde{Z}\|_F) + c_1 + \varepsilon_N, \tag{5.2.32}$$

$$\|w\| \leq C_0 + C_1 \|\tilde{Z}\|_F + C_2 \|\tilde{Z}\|_F \|r\| + C_3 \|\tilde{Z}\|_F \|\tilde{\tau}\| + C_4 \|\tilde{Z}\|_F^2, \tag{5.2.33}$$

where C_i are computable positive constants.

Neural network weight tuning and stability analysis

The next theorem shows how to tune the NN weights so the tracking errors $r(t)$ and $\tilde{\tau}(t)$ achieve small values while the NN weights $\hat{W}$ and $\hat{V}$ are close to W and V so that the weight estimation errors defined by (5.2.24) are bounded.

Theorem 5.2.2 (Control law for backstepping loop). *Let Assumptions 1–3 hold. Select the control input as (5.2.26). Choose the robustifying signal v_2 as*

$$v_2 = K_{Z_1}(\|\hat{Z}\|_F + Z_M)\left(\tilde{\tau} + \|r\|\frac{\tilde{\tau}}{\|\tilde{\tau}\|}\right) + K_{Z_2}\|r\|\frac{\tilde{\tau}}{\|\tilde{\tau}\|} + K_{Z_3}(\|\hat{Z}\|_F + Z_M)^2 \frac{\tilde{\tau}}{\|\tilde{\tau}\|}, \tag{5.2.34}$$

where $K_{Z_1} > \max(C_2, C_3)$, $K_{Z_2} > 1$, and $K_{Z_3} > C_4$. Let the estimated NN weights be provided by the NN tuning algorithm,

$$\dot{\hat{V}} = -T x_{\text{NN}} \tilde{\tau}^T \hat{W}^T \hat{\sigma}' - kT \|\tilde{\tau}\| \hat{V}, \tag{5.2.35}$$

$$\dot{\hat{W}} = -S(\hat{\sigma} - \hat{\sigma}' \hat{V}^T x_{\text{NN}}) \tilde{\tau}^T - kS \|\tilde{\tau}\| \hat{W}, \tag{5.2.36}$$

with any constant matrices $S = S^T > 0$, $T = T^T > 0$, and $k > 0$ small scalar design parameter. Then the filtered tracking error $r(t)$, error $\tau(t)$, and NN weight estimates $\hat{V}$ and $\hat{W}$ are UUB with bounds given by (5.2.37) and (5.2.38). Moreover, the error $\tau(t)$ can be made arbitrarily small by increasing the gain K_b.

Proof. Select the Lyapunov function candidate

$$L = L_1 + \frac{1}{2}\tilde{\tau}^T \tilde{\tau} + \frac{1}{2}\text{tr}(\tilde{W}^T S^{-1} \tilde{W}) + \frac{1}{2}\text{tr}(\tilde{V}^T T^{-1} \tilde{V}),$$

which weights both errors $r(t)$ and $\tau(t)$, and NN weights estimation errors. Taking the derivative

$$\dot{L} = \dot{L}_1 + \tilde{\tau}^T \dot{\tilde{\tau}} + \text{tr}(\tilde{W}^T S^{-1} \dot{\tilde{W}}) + \text{tr}(\tilde{V}^T T^{-1} \dot{\tilde{V}})$$

and using (5.2.8) and (5.2.28), one has

$$\dot{L} = r^T (f(x) + Y_d + \tau_d + \tau)$$
$$+ \tilde{\tau}^T (-K_b \tilde{\tau} - \tilde{W}^T (\hat{\sigma} - \hat{\sigma}' \hat{V}^T x_{\text{NN}}) - \hat{W}^T \hat{\sigma}' \tilde{V}^T x_{\text{NN}} + w - v_2)$$
$$+ \text{tr}(\tilde{W}^T S^{-1} \dot{\tilde{W}}) + \text{tr}(\tilde{V}^T T^{-1} \dot{\tilde{V}}),$$

$$\dot{L} = r^T (f(x) + Y_d + \tau_d + \tau_{\text{des}}) - r^T \tilde{\tau}$$
$$+ \tilde{\tau}^T (-K_b \tilde{\tau} - \tilde{W}^T (\hat{\sigma} - \hat{\sigma}' \hat{V}^T x_{\text{NN}}) - \hat{W}^T \hat{\sigma}' \tilde{V}^T x_{\text{NN}} + w - v_2)$$
$$+ \text{tr}(\tilde{W}^T S^{-1} \dot{\tilde{W}}) + \text{tr}(\tilde{V}^T T^{-1} \dot{\tilde{V}}),$$

$$\dot{L} = r^T (f(x) + Y_d + \tau_d + \tau_{\text{des}}) - r^T \tilde{\tau}$$
$$+ \tilde{\tau}^T (-K_b \tilde{\tau} + w - v_2)$$
$$+ \text{tr}[\tilde{W}^T (S^{-1} \dot{\tilde{W}} - (\hat{\sigma} - \hat{\sigma}' \hat{V}^T x_{\text{NN}}) \tilde{\tau}^T)] + \text{tr}[\tilde{V}^T (T^{-1} \dot{\tilde{V}} - x_{\text{NN}} \tilde{\tau}^T \hat{W}^T \hat{\sigma}')].$$

Applying (5.2.11) and the tuning rules yields

$$\dot{L} = r^T (\tilde{f}(x) + \tau_d - K_V r + v_1) - r^T \tilde{\tau} + \tilde{\tau}^T (-K_b \tilde{\tau} + w - v_2)$$
$$+ k \|\tilde{\tau}\| \text{tr}[\tilde{W}^T (W - \tilde{W})] + k \|\tilde{\tau}\| tr[\tilde{V}^T (V - \tilde{V})].$$

Using the same inequality as in the proof of Theorem 5.2.1, this expression can be bounded as

$$\dot{L} \le -K_{V\min} \|r\|^2 - \|r\|(f_M + \tau_M) + \|r\|\|\tilde{f} + \tau_d\|$$
$$+ k \|\tilde{\tau}\| \|\tilde{Z}\|_F (Z_M - \|\tilde{Z}\|_F) - K_{b\min} \|\tilde{\tau}\|^2 - r^T \tilde{\tau} + \tilde{\tau}^T w$$
$$- \tilde{\tau}^T K_{Z_1} (\|\hat{Z}\|_F + Z_M) \left(\tilde{\tau} + \|r\| \frac{\tilde{\tau}}{\|\tilde{\tau}\|} \right) - \tilde{\tau}^T K_{Z_2} \|r\| \frac{\tilde{\tau}}{\|\tilde{\tau}\|}$$
$$- \tilde{\tau}^T K_{Z_3} (\|\hat{Z}\|_F + Z_M)^2 \frac{\tilde{\tau}}{\|\tilde{\tau}\|}.$$

Including (5.2.33) and applying some norm properties, one has

$$\dot{L} \le -K_{V\min} \|r\|^2 - \|r\|(f_M + \tau_M) + \|r\|\|\tilde{f} + \tau_d\|$$
$$+ k \|\tilde{\tau}\| \|\tilde{Z}\|_F (Z_M - \|\tilde{Z}\|_F) - K_{b\min} \|\tilde{\tau}\|^2 + \|\tilde{\tau}\|\|r\|$$
$$+ C_0 \|\tilde{\tau}\| + C_1 \|\tilde{\tau}\| \|\tilde{Z}\|_F + C_2 \|\tilde{\tau}\| \|\tilde{Z}\|_F \|r\| + C_3 \|\tilde{Z}\|_F \|\tilde{\tau}\|^2 + C_4 \|\tilde{\tau}\| \|\tilde{Z}\|_F^2$$
$$- K_{Z_1} \|\tilde{\tau}\|^2 \|\tilde{Z}\|_F - K_{Z_1} \|r\| \|\tilde{\tau}\| \|\tilde{Z}\|_F - K_{Z_2} \|\tilde{\tau}\| \|r\| - K_{Z_3} \|\tilde{\tau}\| \|\tilde{Z}\|_F^2,$$

$$\dot{L} \le -K_{V\min} \|r\|^2 + k Z_M \|\tilde{\tau}\| \|\tilde{Z}\|_F - k \|\tilde{\tau}\| \|\tilde{Z}\|_F^2 - K_{b\min} \|\tilde{\tau}\|^2 + \|\tilde{\tau}\|\|r\|$$
$$+ C_0 \|\tilde{\tau}\| + C_1 \|\tilde{\tau}\| \|\tilde{Z}\|_F + C_2 \|\tilde{\tau}\| \|\tilde{Z}\|_F \|r\| + C_3 \|\tilde{Z}\|_F \|\tilde{\tau}\|^2 + C_4 \|\tilde{\tau}\| \|\tilde{Z}\|_F^2$$
$$- K_{Z_1} \|\tilde{\tau}\|^2 \|\tilde{Z}\|_F - K_{Z_1} \|r\| \|\tilde{\tau}\| \|\tilde{Z}\|_F - K_{Z_2} \|\tilde{\tau}\| \|r\| - K_{Z_3} \|\tilde{\tau}\| \|\tilde{Z}\|_F^2.$$

Taking $K_{Z_2} > 1$ yields

$$\dot{L} \le -K_{V\min} \|r\|^2 + k Z_M \|\tilde{\tau}\| \|\tilde{Z}\|_F - k \|\tilde{\tau}\| \|\tilde{Z}\|_F^2 - K_{b\min} \|\tilde{\tau}\|^2$$
$$+ C_0 \|\tilde{\tau}\| + C_1 \|\tilde{\tau}\| \|\tilde{Z}\|_F + C_2 \|\tilde{\tau}\| \|\tilde{Z}\|_F \|r\| + C_3 \|\tilde{Z}\|_F \|\tilde{\tau}\|^2 + C_4 \|\tilde{\tau}\| \|\tilde{Z}\|_F^2$$
$$- K_{Z_1} \|\tilde{\tau}\|^2 \|\tilde{Z}\|_F - K_{Z_1} \|r\| \|\tilde{\tau}\| \|\tilde{Z}\|_F - K_{Z_3} \|\tilde{\tau}\| \|\tilde{Z}\|_F^2.$$

Choosing $K_{Z_1} > \max(C_2, C_3)$ and $K_{Z_3} > C_4$, one has

$$\dot{L} \le -K_{V\min} \|r\|^2 + k Z_M \|\tilde{\tau}\| \|\tilde{Z}\|_F - k \|\tilde{\tau}\| \|\tilde{Z}\|_F^2 - K_{b\min} \|\tilde{\tau}\|^2$$
$$+ C_0 \|\tilde{\tau}\| + C_1 \|\tilde{\tau}\| \|\tilde{Z}\|_F,$$

$$\dot{L} \le -K_{V\min} \|r\|^2 - \|\tilde{\tau}\|(K_{b\min} \|\tilde{\tau}\| + k \|\tilde{Z}\|_F^2 - (k Z_M + C_1) \|\tilde{Z}\|_F - C_0).$$

Completing the squares yields

$$\dot{L} \leq -K_{V\min} \|r\|^2$$

$$- \|\tilde{\tau}\| \left[K_{b\min} \|\tilde{\tau}\| + k \left(\|\tilde{Z}\|_F - \left(\frac{Z_M k + C_1}{2k} \right) \right)^2 - k \left(\frac{Z_M k + C_1}{2k} \right)^2 - C_0 \right].$$

Thus $\dot{L}$ is negative as long as either

$$\|\tilde{\tau}\| > \frac{k \left(\frac{Z_M k + C_1}{2k} \right)^2 + C_0}{K_{b\min}} \tag{5.2.37}$$

or

$$\|\tilde{Z}\|_F > \frac{Z_M k + C_1}{2k} + \sqrt{\left(\frac{Z_M k + C_1}{2k} \right)^2 + \frac{C_0}{k}}. \qquad \Box \tag{5.2.38}$$

Some remarks will emphasize the important aspects of our NN backlash compensator. Note first that, in practice, it is not necessary to compute the bounding constants C_2, C_3, and C_4 mentioned in the theorem statement since one may simply select the gains K_{Z_i} large. This is accomplished using computer simulation to verify their suitability.

Unsupervised backpropagation through time with extra terms

The first terms of (5.2.35) and (5.2.36) are modified versions of the standard backpropagation algorithm. The k terms correspond to the e-mod (Narendra and Annaswamy (1987)), which guarantees bounded NN weight estimates. Note that the robustifying term consists of three terms. The first and third terms are specifically designed to ensure the stability of the overall system in the presence of the disturbance term (5.2.29). The second term is included to ensure stability in the presence of the error $\tilde{\tau} = \tau_{\text{des}} - \tau$ in the backstepping design.

Bounds on the tracking error and neural network weight estimation errors

The right-hand side of (5.2.37) can be taken as a practical bound on the error in the sense that $\tilde{\tau}(t)$ will never stay far above it. Note that the stability radius may be decreased any amount by increasing the gain K_b. It is noted that PD control without backlash compensation requires much higher gains in order to achieve similar performance—that is, eliminating the NN feedforward compensator will result in degraded performance. Moreover, it is difficult to prove the stability of such a highly nonlinear system using only PD. Using NN backlash compensation, the stability of the system is formally proven and the tracking error can be kept arbitrarily small by increasing the gain K_b.

On the other hand, (5.2.38) reveals that the NN weight errors cannot be made arbitrarily small. In fact, they are fundamentally bounded in terms of V_M and W_M (which are contained in Z_M). We are not concerned about this as practical engineers as long as the NN weight estimation errors are bounded, which is required to ensure that the control signal (5.2.26) is bounded.

Neural network weight initialization

Many other NN control techniques in the literature require one to find initial stabilizing weights for the NN controller. This is well nigh impossible. In our NN backlash compensator, on the other hand, the weights are easy to initialize. Note, in fact, that if the NN weights are all set to zero, the NN is effectively absent in Figure 5.2.3. Then the feedforward path with gains K_b allows the PD control tracking loop to stabilize the plant until the NN begins to learn to compensate for the backlash. This structure is a result of the decomposition performed in Figure 5.2.2. However, one cannot set all the NN weights to zero: by examining the tuning laws, we see that the weights may remain at zero forever. In practice, the weights V are initialized to random values. It is shown in Igelnik and Pao (1995) that this random variable functional link NN provides a basis for function approximation. The weights W are initialized at zero.

Linearity in the model parameters

The proposed backlash compensator does not require LIP. The standard adaptive control techniques for backlash compensation usually require this assumption (Tao and Kokotović (1996)). The LIP requirement is a severe restriction for practical systems, and moreover it is necessary to find the linear parametrization, which can be tedious. Instead, we use here the NN approximation property, which holds over a compact set.

5.2.4   Simulation of neural network backlash compensator

To illustrate the performance of the NN backlash compensator, we consider the nonlinear system

$$\dot{x}_1 = x_2,$$
$$\dot{x}_2 = -\frac{1}{T}x_2 + max_2^2 \sin(x_1) + mga\cos(x_1) + \tau, \qquad (5.2.39)$$

which represents the mechanical motion of simple robot-like systems. The motor time constant is T, m is a net effective load mass, a is a length, and g is the gravitational constant. We selected $T = 1$s, $m = 1$kg, and $a = 2.5$m. The input τ has been passed through the backlash nonlinearity given by (5.2.1). The parameters of the backlash are chosen as $d_+ = 10$, $d_- = -10.5$, and $m = 1$. The desired trajectory is selected as $\sin(t)$.

The NN weight tuning parameters are chosen as $S = 30I_{11}$, $T = 50I_6$, and $k = 0.01$, where I_N is the $N \times N$ identity matrix. The robustifying signal gains are $K_{z1} = 2$, $K_{z2} = 2$, and $K_{z3} = 2$. The controller parameters are chosen as $\Lambda = 5$, $K_v = 10$, and $K_b = 15$.

The NN has L=10 hidden-layer nodes with sigmoidal activation functions. The first-layer weights V are initialized randomly (Igelnik and Pao (1995)). They are uniformly randomly distributed between -1 and 1. The second-layer weights W are initialized at zero. Note that this weight initialization will not affect system stability since the NN throughput is initially zero, and therefore there is initially no input to the system except for the PD loop and the inner feedback linearization loop. The filter that generates the signal $\dot{\tau}_{\text{des}}$ is implemented as $\frac{s}{s+100}$. This approximates the derivative required in backstepping.

The NN parameters and the number of hidden-layer neurons were selected by performing several simulations with different parameter values until good performance was obtained. With our software, each simulation takes only a few minutes. One selected $L = 10$ because the performance was degraded using $L = 8$ and there was no further improvement in selecting $L = 12$.

We simulated three cases. Figures 5.2.4 and 5.2.5 show the tracking performance of the closed-loop system with only a PD controller, which amounts to $\hat{\varphi} = K_b\tilde{\tau} + \dot{\tau}_{\text{des}}$. It can be seen that backlash degrades the system performance.

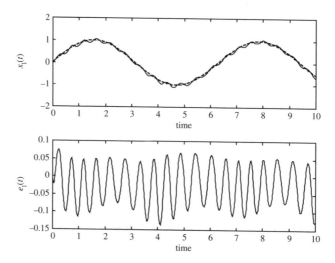

Figure 5.2.4. *State $x_1(t)$ and tracking error $e_1(t)$, $\hat{\varphi} = K_b\tilde{\tau} + \dot{\tau}_{des}$.*

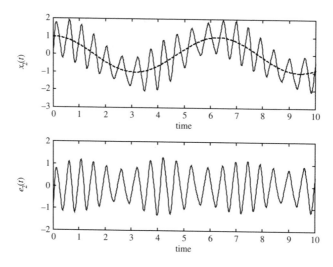

Figure 5.2.5. *State $x_2(t)$ and tracking error $e_2(t)$, $\hat{\varphi} = K_b\tilde{\tau} + \dot{\tau}_{des}$.*

In the next case, we simulated the same nonlinear system with the same backlash, but now using $\hat{\varphi} = K_b\tilde{\tau} + \dot{\tau}_{des} + v_2$. In this case, the backlash compensator consists of the PD loop plus an additional robustifying term. The results are shown in Figures 5.2.6 and 5.2.7. They are significantly improved over the case of using only PD control.

In the last case, we simulated the complete backlash compensator, as is shown in Figure 5.2.3. The NN is now included in the feedforward path, and one has $\hat{\varphi} = K_b\tilde{\tau} + \dot{\tau}_{des} - \hat{W}^T\sigma(\hat{V}^T x_{NN}) + v_2$. Figures 5.2.8 and 5.2.9 show the tracking errors when the NN compensator is included. One can see that after 0.5 seconds the NN has adjusted its weights online such that the backlash effect is greatly reduced. Applying the NN backlash compensator greatly reduces the tracking error.

Some of the NN weights are shown in Figure 5.2.10. One can see that the NN weights W vary through time, trying to compensate for the system nonlinearity, while the NN weights

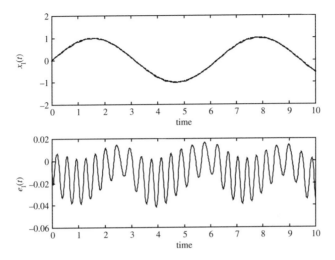

Figure 5.2.6. *State $x_1(t)$ and tracking error $e_1(t)$, $\hat{\varphi} = K_b \tilde{\tau} + \dot{\tau}_{\text{des}} + v_2$.*

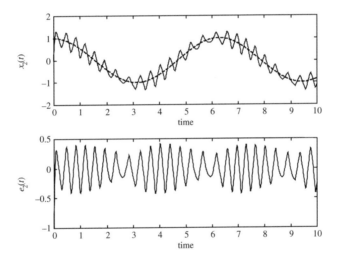

Figure 5.2.7. *State $x_2(t)$ and tracking error $e_2(t)$, $\hat{\varphi} = K_b \tilde{\tau} + \dot{\tau}_{\text{des}} + v_2$.*

V, which determine the positions and shapes of the sigmoid functions, vary through time more slowly.

From this simulation it is clear that the proposed NN backlash compensator is an efficient way to compensate for backlash nonlinearities of all kinds, without any restrictive assumptions on the backlash model itself.

5.3 Discrete-time neural network backlash compensation

Many systems with actuator nonlinearities such as deadzone and backlash are modeled in discrete time. An example of deadzone in biomedical control is the functional neuro-muscular stimulation for restoring motor function by directly activating paralyzed muscles

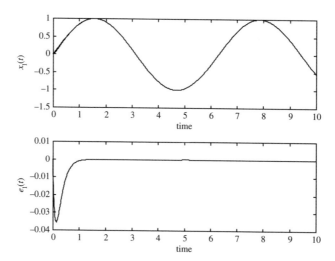

Figure 5.2.8. *State $x_1(t)$ and tracking error $e_1(t)$, $\hat{\varphi} = K_b\tilde{\tau} + \dot{\tau}_{\text{des}} - \hat{W}^T\sigma(\hat{V}^T x_{\text{NN}}) + v_2$.*

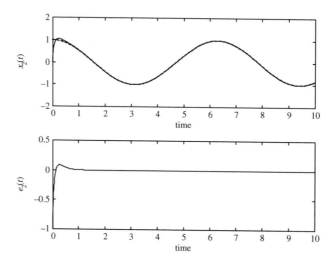

Figure 5.2.9. *State $x_2(t)$ and tracking error $e_2(t)$, $\hat{\varphi} = K_b\tilde{\tau} + \dot{\tau}_{\text{des}} - \hat{W}^T\sigma(\hat{V}^T x_{\text{NN}}) + v_2$.*

(Bernotas, Crago, and Chizeck (1987)). Moreover, for implementation as a digital controller, a discrete-time actuator nonlinearity compensator is needed. A discrete-time FL deadzone compensation scheme was proposed in Chapter 4. To address discrete-time deadzone compensation, an adaptive control approach has been proposed in the seminal work by Tao and Kokotović (1994, 1996). Adaptive control approaches for backlash compensation in discrete time were presented in Grundelius and Angelli (1996) and Tao and Kokotović (1994, 1996). These all require an LIP assumption, which may not hold for actual industrial motion systems.

Since backlash is a dynamic nonlinearity, a backlash compensator should also be dynamic. We intend to use discrete-time dynamic inversion in this section to design a discrete-time backlash compensator. Dynamic inversion is a form of backstepping. Backstepping was

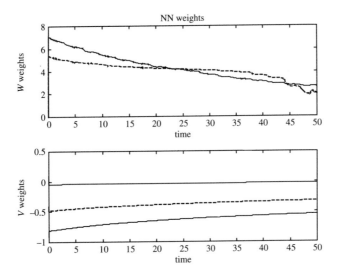

Figure 5.2.10. *Some of the NN weights.*

extended to discrete-time systems in Byrnes and Lin (1994), Haddad, Fausz, and Abdallah (1998), Jagannathan (1998), Yeh and Kokotović (1995), and Zhang and Soh (1998). The difficulty with discrete-time dynamic inversion is that a future value of a certain ideal control input is needed. This chapter shows how to confront that problem for backlash compensation by adding a filter.

In this section, we provide the complete solution for extending dynamic inversion to discrete-time systems by using a filtered prediction approach for backlash compensation. The general case of nonsymmetric backlash is treated. A rigorous design procedure is given that results in a PD tracking loop with an adaptive NN in the feedforward loop for dynamic inversion of the backlash nonlinearity. The NN feedforward compensator is adapted in such a way as to estimate online the backlash inverse. Unlike standard discrete-time adaptive control techniques, no CE assumption is needed since the tracking error and the estimation error are weighted in *the same* Lyapunov function. No LIP is needed. The approach is similar to that in Jagannathan and Lewis (1996), but additional complexities arise due to the fact that the backlash compensator is in the *feedforward* loop.

5.3.1 Discrete-time backlash nonlinearity and backlash inverse

The backlash nonlinearity is shown in Figure 5.1.1, and the mathematical model for the discrete-time case is

$$\tau(k+1) = B(\tau(k), u(k), u(k+1))$$

$$= \begin{cases} mu(k+1) & \text{if } u(k) = m(\tau(k) - d_+), \\ & \text{if } u(k+1) > 0 \text{ and } u(k) = m(\tau(k) - d_-), \\ 0 & \text{otherwise.} \end{cases} \tag{5.3.1}$$

It can be seen that backlash is a first-order velocity-driven dynamic system with inputs $u(k)$ and $u(k+1)$ and state $\tau(k)$. Backlash contains its own dynamics; therefore, its compensation requires the design of a dynamic compensator.

Whenever the motion $u(k)$ changes its direction, the motion $\tau(k)$ is delayed from motion of $u(k)$. The objective of the backlash compensator is to make this delay as small as possible (i.e., to make the throughput from $u(k)$ to $\tau(k)$ be the unity). The backlash precompensator needs to generate the inverse of the backlash nonlinearity. The backlash inverse function is shown in Figure 5.2.1. The dynamics of the NN backlash compensator may be written in the form

$$u(k + 1) = B_{\text{inv}}(u(k), w(k), w(k + 1)). \tag{5.3.2}$$

The backlash inverse characteristic shown in Figure 5.2.1 can be decomposed into two functions: a known direct feedforward term plus an additional parallel path containing a *modified backlash inverse* term as shown in Figure 5.2.2. This decomposition allows the design of a compensator that has a better structure than when a backlash compensator is used directly in the feedforward path.

5.3.2   Dynamics and control of a nonlinear discrete-time system

Consider an mnth-order MIMO discrete-time system given by

$$x_1(k + 1) = x_2(k),$$

$$\vdots \tag{5.3.3}$$

$$x_{n-1}(k + 1) = x_n(k),$$
$$x_n(k + 1) = f(x(k)) + \tau(k) + d(k),$$

where $x(k) = [x_1(k), x_2(k), \ldots, x_n(k)]^T$, with $x_i(k) \in \Re^n$, $i = 1, 2, \ldots, n$, and $\tau(k) \in R^m$. The vector $d(k) \in R^m$ denotes an unknown disturbance acting on the system at the instant k with $\|d(k)\| \le d_M$ a known constant. The actuator output $\tau(k)$ is related to the control input $u(k)$ through the backlash nonlinearity $\tau(k) = \text{Backlash}(u(k))$.

Given a desired trajectory $x_{nd}(k)$ and its delayed values, define the tracking error as

$$e_n(k) = x_n(k) - x_{nd}(k). \tag{5.3.4}$$

It is typical in robotics to define a so-called filtered tracking error $r(k) \in R^m$ given by

$$r(k) = e_n(k) + \lambda_1 e_{n-1}(k) + \cdots + \lambda_{n-1} e_1(k), \tag{5.3.5}$$

where $e_{n-1}(k), \ldots, e_1(k)$ are the delayed values of the error $e_n(k)$ and $\lambda_1, \ldots, \lambda_{n-1}$ are constant matrices selected so that $|z^{n-1} + \lambda_1 z^{n-2} + \cdots + \lambda_{n-1}|$ is stable or Hurwitz (i.e., $e_n(k) \to 0$ exponentially as $r(k) \to 0$). This equation can be further expressed as

$$r(k + 1) = e_n(k + 1) + \lambda_1 e_{n-1}(k + 1) + \cdots + \lambda_{n-1} e_1(k + 1). \tag{5.3.6}$$

Using (5.3.3) in (5.3.6), the dynamics of the mnth-order MIMO system can be written in terms of the tracking error as

$$r(k + 1) = f(x(k)) - x_{nd}(k + 1) + \lambda_1 e_{n-1}(k) + \cdots + \lambda_{n-1} e_1(k) + \tau(k) + d(k). \tag{5.3.7}$$

The following standard assumptions are needed, and they are true in every practical situation.

Assumption 4 (Bounded disturbance). The unknown disturbance satisfies $\|d(k)\| \leq d_M$ with d_M a known positive constant.

Assumption 5 (Bounded estimation error). The nonlinear function is assumed to be unknown, but a fixed estimate $\hat{f}(x(k))$ is assumed known such that the functional estimation error $\tilde{f}(x(k)) = f(x(k)) - \hat{f}(x(k))$ satisfies $\|\tilde{f}(x(k))\| \leq f_M(x(k))$ for some known bounding function $f_M(x(k))$.

Assumption 6 (Bounded desired trajectories). The desired trajectory is bounded in the sense that

$$\left\|\begin{array}{c} x_{1d}(k) \\ x_{2d}(k) \\ \vdots \\ x_{nd}(k) \end{array}\right\| \leq X_d \qquad (5.3.8)$$

for a known bound X_d.

Ideal control law assuming no backlash

The discrete-time NN backlash compensator is to be designed using the backstepping technique. First, we will design a compensator that guarantees system trajectory tracking when there is no backlash. A robust compensation scheme for unknown terms in $f(x(k))$ is provided by selecting the tracking controller

$$\tau_{\text{des}}(k) = K_v \cdot r(k) - \hat{f}(x(k)) + x_{nd}(k+1) - \lambda_1 \cdot e_{n-1}(k) - \lambda_2 \cdot e_{n-2}(k) \\ - \cdots - \lambda_{n-1} \cdot e_1(k), \qquad (5.3.9)$$

with $\hat{f}(x(k))$ an estimate for the nonlinear terms $f(x(k))$. The feedback gain matrix $K_v > 0$ is often selected diagonal. The problem of finding $\hat{f}(x(k))$ is not the main concern of this chapter, so it is considered to be available. This is a robust control approach given a fixed estimate for $f(x)$. We assume that the estimation error is bounded so that Assumption 5 holds.

Using (5.3.9) as a control input, the system dynamics in (5.3.7) can be rewritten as

$$r(k+1) = K_v \cdot r(k) + \tilde{f}(x(k)) + d(k). \qquad (5.3.10)$$

The next theorem is the first step in the backstepping design, and it shows that if there is no backlash, the desired control law (5.3.9) will keep the filtered tracking error small in the sense of uniform ultimate boundedness (Chapter 2).

Theorem 5.3.1 (Control law for outer tracking loop). *Consider the system given by (5.3.3). Assume that Assumptions 4 and 5 hold, and let the control action be provided by (5.3.9) with $0 < K_v < I$ being a design parameter. Then the filtered tracking error $r(k)$ is UUB.*

Proof. Let us consider the Lyapunov function candidate

$$L_1(k) = r(k)^T r(k).$$

The first difference is

$$\Delta L_1(k) = r(k+1)^T r(k+1) - r(k)^T r(k) \\ = (K_v \cdot r(k) + \tilde{f}(x(k)) + d(k))^T (K_v \cdot r(k) + \tilde{f}(x(k)) + d(k)) - r(k)^T r(k).$$

$\Delta L_1(k)$ is negative if $\|K_v r(k) + \tilde{f}(x(k)) + d(k)\| \leq K_{v\,\text{max}} \|r(k)\| + f_M + d_M < \|r(k)\|$, which implies that $(1 - K_{v\,\text{max}}) \|r(k)\| > f_M + d_M$, which is true as long as

$$\|r(k)\| > \frac{f_M + d_M}{1 - K_{v\,\text{max}}}.$$

Therefore, $\Delta L_1(k)$ is negative outside a compact set. According to a standard Lyapunov theory extension, this demonstrates the UUB of $r(k)$. $\square$

5.3.3   Discrete-time neural network backlash compensation using dynamic inversion

The discrete-time NN backlash compensator is to be designed using the backstepping technique. It is assumed that the actuator output $\tau(k)$ is measurable. Backstepping design for backlash compensation has two steps. First, one designs an ideal control law that gives good performance if there is no backlash. This control law is given in Theorem 5.3.1. Unfortunately, in the presence of an unknown backlash nonlinearity, the desired and actual value of the control signal $\tau(k)$ will be different. In this situation, a dynamics inversion technique by NNs can be used for compensation of the inversion error. This is derived by using the second step in backstepping, which we now perform. The objective is to make $\tau(k)$ closely follow $\tau_{\text{des}}(k)$.

The actuator output given by (5.3.9) is the desired control signal. The complete error system dynamics can be found defining the error

$$\tilde{\tau}(k) = \tau_{\text{des}}(k) - \tau(k). \tag{5.3.11}$$

Using the desired control input (5.3.9), under the presence of unknown backlash, the system dynamics (5.3.7) can be rewritten as

$$r(k+1) = K_v \cdot r(k) + \tilde{f}(x(k)) + d(k) - \tilde{\tau}(k). \tag{5.3.12}$$

Evaluating (5.3.11) at the subsequent time interval,

$$\tilde{\tau}(k+1) = \tau_{\text{des}}(k+1) - \tau(k+1) = \tau_{\text{des}}(k+1) - B(\tau(k), u(k), u(k+1)), \tag{5.3.13}$$

which together with (5.3.12) represents the complete system error dynamics.

The dynamics of the backlash nonlinearity can be written as

$$\tau(k+1) = \varphi(k), \tag{5.3.14}$$
$$\varphi(k) = B(\tau(k), u(k), u(k+1)), \tag{5.3.15}$$

where $\varphi(k)$ is a pseudocontrol input (Leitner, Calise, and Prasad (1997)). In the case of known backlash, the ideal backlash inverse is given by

$$u(k+1) = B^{-1}(u(k), \tau(k), \varphi(k)). \tag{5.3.16}$$

Since the backlash and therefore its inverse are not known, one can only approximate the backlash inverse as

$$\hat{u}(k+1) = \hat{B}^{-1}(\hat{u}(k), \tau(k), \hat{\varphi}(k)). \tag{5.3.17}$$

The backlash dynamics can now be written as

$$\tau(k+1) = B(\tau(k), \hat{u}(k), \hat{u}(k+1))$$
$$= \hat{B}(\tau(k), \hat{u}(k), \hat{u}(k+1)) + \tilde{B}(\tau(k), \hat{u}(k), \hat{u}(k+1)) \qquad (5.3.18)$$
$$= \hat{\varphi}(k) + \tilde{B}(\tau(k), \hat{u}(k), \hat{u}(k+1)),$$

where $\hat{\varphi}(k) = \hat{B}(\tau(k), \hat{u}(k), \hat{u}(k+1))$, and therefore its inverse is given by $\hat{u}(k+1) = \hat{B}^{-1}(\tau(k), \hat{u}(k), \hat{\varphi}(k))$. The unknown function $\tilde{B}(\tau(k), \hat{u}(k), \hat{u}(k+1))$, which represents the backlash inversion error, will be approximated using an NN.

In order to design a stable closed-loop system with backlash compensation, one selects a nominal backlash inverse $\hat{u}(k+1) = \hat{\varphi}(k)$ and pseudocontrol input as

$$\hat{\varphi}(k) = -K_b \tilde{\tau}(k) + \tau_{\text{filt}}(k) + \hat{W}(k)^T \sigma(V^T x_{\text{NN}}(k)), \qquad (5.3.19)$$

where $K_b > 0$ is a design parameter and τ_{filt} is a discrete-time filtered version of τ_{des}. τ_{filt} is a filtered prediction that approximates $\tau_{\text{des}}(k+1)$ and is obtained using the discrete-time filter $\frac{az}{z+a}$ as shown in Figure 5.3.1. This is the equivalent of using a filtered derivative instead of a pure derivative in continuous-time dynamics inversion, as required in industrial control systems. The filter dynamics shown in Figure 5.3.1 can be written as

$$\tau_{\text{filt}}(k) = -\frac{\tau_{\text{filt}}(k+1)}{a} + \tau_{\text{des}}(k+1), \qquad (5.3.20)$$

where a is a design parameter. It can be seen that when the filter parameter a is large enough, we have $\tau_{\text{filt}}(k) \approx \tau_{\text{des}}(k+1)$. The mismatch term $(-\tau_{\text{filt}}(k+1)/a)$ can be approximated along with the backlash inversion error using the NN.

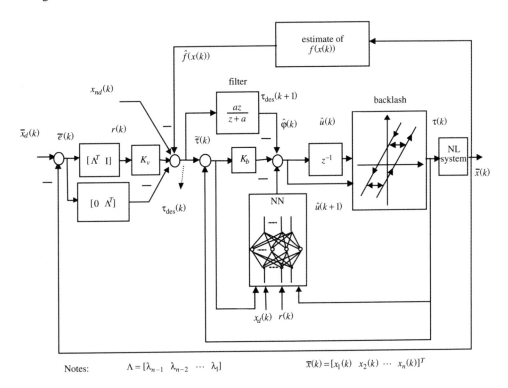

Figure 5.3.1. *Discrete-time NN backlash compensator.*

Based on the NN approximation property, the backlash inversion plus the filtered error dynamics can be represented as

$$\tilde{B}(\tau(k), \hat{u}(k), \hat{u}(k+1)) + \frac{\tau_{\text{filt}}(k+1)}{a} = W(k)^T \sigma(V^T x_{\text{NN}}(k)) + \varepsilon(k), \qquad (5.3.21)$$

where the NN input vector is chosen to be $x_{\text{NN}}(k) = [1 \quad r(k)^T \quad x_d(k)^T \quad \tilde{\tau}(k)^T \quad \tau(k)^T]^T$ and $\varepsilon(k)$ represents the NN approximation error. Notice that the first-layer weights V are not time dependent since they are selected randomly at the initial time to provide a basis (Igelnik and Pao (1995)) and then they are kept constant through the tuning process.

Define the NN weight estimation error as

$$\tilde{W}(k) = W(k) - \hat{W}(k), \qquad (5.3.22)$$

where $\hat{W}(k)$ is the estimate of the ideal NN weights $W(k)$.

Using the proposed controller shown in Figure 5.3.1, the error dynamics can be written as

$$
\begin{aligned}
\tilde{\tau}(k+1) &= \tau_{\text{des}}(k+1) - \hat{\varphi}(k) + \tilde{B}(\tau(k), \hat{u}(k), \hat{u}(k+1)) \\
&= K_b \tilde{\tau}(k) + \frac{\tau_{\text{filt}}(k+1)}{a} - \hat{W}(k)^T \sigma(V^T x_{\text{NN}}(k)) + \tilde{B}(\tau(k), \hat{u}(k), \hat{u}(k+1)) \\
&= K_b \tilde{\tau}(k) - \hat{W}(k)^T \sigma(V^T x_{\text{NN}}(k)) + W(k)^T \sigma(V^T x_{\text{NN}}(k)) + \varepsilon(k).
\end{aligned}
$$
$$(5.3.23)$$

Using (5.3.22),

$$\tilde{\tau}(k+1) = K_b \tilde{\tau}(k) + \tilde{W}(k)^T \sigma(V^T x_{\text{NN}}(k)) + \varepsilon(k). \qquad (5.3.24)$$

The next theorem is our main result and shows how to tune the NN weights so the tracking error $r(k)$ and backlash estimation error $\tilde{\tau}(k)$ achieve small values, while the NN weights estimation errors $\tilde{W}(k)$ are bounded.

Theorem 5.3.2 (Control law for backstepping loop). *Consider the system given by (5.3.3). Provided that Assumptions 4–6 hold, let the control action $\hat{\varphi}(k)$ be provided by (5.3.19), with $K_b > 0$ being a design parameter. Let the control input be provided by $\hat{u}(k+1) = \hat{\varphi}(k)$ and the estimated NN weights be provided by the NN tuning law*

$$\hat{W}(k+1) = \hat{W}(k) + \alpha\sigma(k)r(k+1)^T + \alpha\sigma(k)\tilde{\tau}(k+1)^T - \Gamma \|I - \alpha\sigma(k)\sigma(k)^T\| \hat{W}(k), \qquad (5.3.25)$$

where $\alpha > 0$ is a constant learning rate parameter or adaptation gain, $\Gamma > 0$ is a design parameter, and for simplicity purposes $\sigma(V^T x_{\text{NN}}(k))$ is represented as $\sigma(k)$. Then the filtered tracking error $r(k)$, the backlash estimation error $\tilde{\tau}(k)$, and the NN weight estimation error $\tilde{W}(k)$ are UUB provided the following conditions hold:

$$(1) \quad 0 < \alpha\sigma(k)^T\sigma(k) < \frac{1}{2}, \qquad (5.3.26)$$

$$(2) \quad 0 < \Gamma < 1, \qquad (5.3.27)$$

$$(3) \quad 0 < K_v < I \quad and \quad K_{v\max} < \frac{1}{\sqrt{\eta+2}}, \qquad (5.3.28)$$

where

$$\beta = K_v^{-1}(2I - K_v) + (1 - \alpha\sigma(k)^T\sigma(k))K_v^{-T}K_v^{-1} > 0, \tag{5.3.29}$$

$$\rho = (1 - \alpha\sigma(k)^T\sigma(k))I - \beta^{-1}(\alpha\sigma(k)^T\sigma(k) + \Gamma\|I - \alpha\sigma(k)\sigma(k)^T\|)^2 > 0, \tag{5.3.30}$$

$$\eta = (1 + \alpha\sigma(k)^T\sigma(k))I + \rho^{-1}(\alpha\sigma(k)^T\sigma(k) + \Gamma\|I - \alpha\sigma(k)\sigma(k)^T\|)^2 > 0. \tag{5.3.31}$$

Note that condition (5.3.29) is true because of (5.3.26) and (5.3.27). Note also that (5.3.31) is satisfied because of conditions (5.3.26) and (5.3.30).

***Proof of condition* (5.3.30).** Because of condition (5.3.26), we have that $(1 - \alpha\sigma^T\sigma)I > \frac{1}{2}I$. Also using (5.3.26) and (5.3.27), we have that

$$(\alpha\sigma^T\sigma + \Gamma\|I - \alpha\sigma\sigma^T\|)^2 < \frac{1}{4}I.$$

Using (5.3.28), we have that $\beta > I$ (i.e., $\beta^{-1} < I$). Then we can conclude that $\beta^{-1}(\alpha\sigma^T\sigma + \Gamma\|I - \alpha\sigma\sigma^T\|)^2 < \frac{1}{4}I$. Finally, using this last result, we can show that

$$\rho = (1 - \alpha\sigma^T\sigma)I - \beta^{-1}(\alpha\sigma^T\sigma + \Gamma\|I - \alpha\sigma\sigma^T\|)^2 > \frac{1}{4}I > 0. \quad \square$$

***Proof of Theorem* 5.3.2.** See the appendix (section 5.A) of this chapter.

It is important to note that in this theorem there is no CE assumption, in contrast to standard work in discrete-time adaptive control. In the latter, a parameter identifier is first selected and the parameter estimation errors are assumed small. Then in the tracking proof, it is assumed that the parameter estimates are exact (the CE assumption), and a Lyapunov function is selected that weights only the tracking error to demonstrate closed-loop stability and tracking performance. By contrast, in our proof, the Lyapunov function shown in the appendix (section 5.A) of this chapter is of the form

$$J(k) = [r(k) + \tilde{\tau}(k)]^T \cdot [r(k) + \tilde{\tau}(k)] + r(k)^T r(k) + \frac{1}{\alpha}\text{tr}\{\tilde{W}(k)^T \cdot \tilde{W}(k)\} > 0, \tag{5.3.32}$$

which weights the tracking error $r(k)$, the backlash estimation error $\tilde{\tau}(k)$, and the NN weight estimation error $\tilde{W}(k)$. This requires an exceedingly complex proof in which we must complete the squares several times with respect to different variables. However, it obviates the need for any sort of CE assumption. It also allows the parameter-tuning algorithm to be derived during the proof process, not selected a priori in an ad hoc manner. Note that additional complexities arise in the proof due to the fact that the backlash compensator NN system is in the *feedforward* loop, not the feedback loop.

The third term in (5.3.25) is a discrete-time version of Narendra's *e*-mod (Narendra and Annaswamy (1987)), which is required to provide robustness due to the coupling in the proof between tracking error, backlash error terms, and weight estimation error terms in the Lyapunov function. This is called a "forgetting term" in NN weight-tuning algorithms. It is required in that context to prevent parameter overtraining.

5.3.4   Simulation results

Now, the discrete-time NN backlash compensator is simulated on a digital computer. It is found to be very efficient at canceling the deleterious effects of actuator backlash. We simulate the response for the known plant with input backlash, both with and without the NN compensator. The C code used in this simulation is contained in Appendix C of the book.

Consider the following nonlinear plant:

$$x_1(k+1) = x_2(k),$$
$$x_2(k+1) = -\frac{3}{16}\left[\frac{x_1(k)}{1+x_2^2(k)}\right] + x_2(k) + \tau(k). \tag{5.3.33}$$

The process input $\tau(k)$ is related to the control input $u(k)$ through the backlash nonlinearity. The deadband widths for the backlash nonlinearity were selected as $d_+ = d_- = 0.2$ and the slope as $m = 0.5$.

We simulate the trajectory tracking performance of the system for sinusoidal reference signals. The reference signal used was selected to be

$$x_d(k) = \sin(w \cdot t_k + \phi), \quad w = 0.5, \quad \phi = \frac{\pi}{2}. \tag{5.3.34}$$

The sampling period was selected as $T = 0.001$ seconds.

Figure 5.3.2 shows the system response without backlash using a standard PD controller. The PD controller does a good job on the tracking, which is achieved at about 2 seconds. Figure 5.3.3 shows the system response using only a PD controller with input backlash. The system backlash destroys the tracking, and the PD controller by itself is not capable of compensating for it.

Now we added the NN backlash compensator prescribed in Theorem 5.3.2. See the discussion in the simulation of section 5.2.4 on selecting the NN parameters. Figure 5.3.4 shows the results using the discrete-time NN backlash compensator. The backlash compensator takes care of the system backlash and the tracking is achieved in less than 0.5 seconds. Thus the NN compensator significantly improves the performance of this system in the presence of backlash.

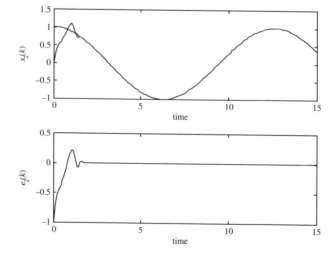

Figure 5.3.2. *PD controller without backlash.*

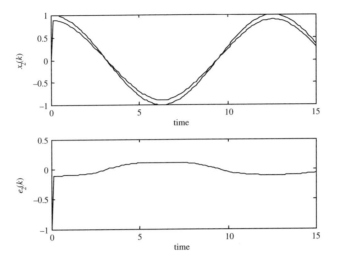

Figure 5.3.3. *PD controller with backlash.*

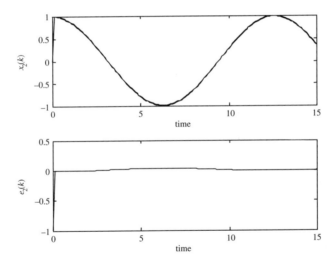

Figure 5.3.4. *PD controller with NN backlash compensation.*

5.A   Appendix: Proof of Theorem 5.3.2

Note: For purposes of simplicity, from now on we will omit the k subindex. Thus every variable is supposed to have a k subindex unless specified otherwise. This statement is valid only for the proofs shown in the appendices.

For purposes of simplicity, let us rewrite the system dynamics as

$$r_{k+1} = K_v \cdot r + D - \tilde{\tau},$$

where $D = \tilde{f} + d$. And let us rewrite the backlash dynamics as

$$\tilde{\tau}_{k+1} = K_b \tilde{\tau} + \tilde{W}^T \sigma (V^T x_{\text{NN}}) + \varepsilon.$$

Select the Lyapunov function candidate

$$L = [r^T \quad \tilde{\tau}^T] \begin{bmatrix} 2 & 1 \\ 1 & 1 \end{bmatrix} \begin{bmatrix} r \\ \tilde{\tau} \end{bmatrix} + \frac{1}{\alpha} \text{tr}(\tilde{W}^T \tilde{W}) > 0.$$

This can be rewritten as

$$L = 2r^T r + 2r^T \tilde{\tau} + \tilde{\tau}^T \tilde{\tau} + \frac{1}{\alpha} \text{tr}(\tilde{W}^T \tilde{W}) = L_1 + L_2 + L_3 + L_4.$$

Taking the first difference,

$$\Delta L_1 = 2r_{k+1}^T r_{k+1} - 2r^T r = 2[K_v r + D - \tilde{\tau}]^T [K_v r + D - \tilde{\tau}] - 2r^T r$$
$$= -2r^T[I - K_v^T K_v]r + 4r^T K_v^T D - 4r^T K_v^T \tilde{\tau} + 2D^T D - 4D^T \tilde{\tau} + 2\tilde{\tau}^T \tilde{\tau},$$

$$\Delta L_2 = 2r_{k+1}^T \tilde{\tau}_{k+1} - 2r^T \tilde{\tau} = 2(K_v r + D - \tilde{\tau})^T (K_b \tilde{\tau} + \tilde{W}^T \sigma + \varepsilon) - 2r^T \tilde{\tau}$$
$$= 2r^T K_v^T K_b \tilde{\tau} + 2r^T K_v^T \tilde{W}^T \sigma + 2r^T K_v^T \varepsilon + 2D^T K_b \tilde{\tau} + 2D^T \tilde{W}^T \sigma$$
$$+ 2D^T \varepsilon - 2\tilde{\tau}^T K_b^T K_b \tilde{\tau} - 2\tilde{\tau}^T \tilde{W}^T \sigma - 2\tilde{\tau}^T \varepsilon - 2r^T \tilde{\tau},$$

$$\Delta L_3 = \tilde{\tau}_{k+1}^T \tilde{\tau}_{k+1} - \tilde{\tau}^T \tilde{\tau} = (K_b \tilde{\tau} + \tilde{W}^T \sigma + \varepsilon)^T (K_b \tilde{\tau} + \tilde{W}^T \sigma + \varepsilon) - \tilde{\tau}^T \tilde{\tau}$$
$$= \tilde{\tau}^T K_b^T K_b \tilde{\tau} + 2\tilde{\tau}^T K_b^T \tilde{W}^T \sigma + 2\tilde{\tau}^T K_b^T \varepsilon + \sigma^T \tilde{W} \tilde{W}^T \sigma + 2\varepsilon^T \tilde{W}^T \sigma + \varepsilon^T \varepsilon - \tilde{\tau}^T \tilde{\tau},$$

$$\Delta L_4 = \frac{1}{\alpha} \text{tr}(\tilde{W}_{k+1}^T \tilde{W}_{k+1} - \tilde{W}^T \tilde{W}) = \frac{1}{\alpha} \text{tr}((W_{k+1} - \hat{W}_{k+1})^T (W_{k+1} - \hat{W}_{k+1}) - \tilde{W}^T \tilde{W})$$
$$= \frac{1}{\alpha} \text{tr}(\hat{W}_{k+1}^T \hat{W}_{k+1} + W^T W - 2W^T \hat{W}_{k+1} - \tilde{W}^T \tilde{W}).$$

Select the tuning law

$$\hat{W}_{k+1} = \hat{W} + \alpha \sigma \cdot r_{k+1}^T + \alpha \sigma \cdot \tilde{\tau}_{k+1}^T - \Gamma \|I - \alpha \cdot \sigma \cdot \sigma^T\| \hat{W}.$$

Then

$$\Delta L_4 = \frac{1}{\alpha} \text{tr}\{[\hat{W} + \alpha \sigma (r_{k+1}^T + \tilde{\tau}_{k+1}^T) - \Gamma \|I - \alpha \sigma \sigma^T\| \hat{W}]^T [\hat{W} + \alpha \sigma (r_{k+1}^T + \tilde{\tau}_{k+1}^T)$$
$$- \Gamma \|I - \alpha \sigma \sigma^T\| \hat{W}]$$
$$+ W^T W - \tilde{W}^T \tilde{W} - 2W^T [\hat{W} + \alpha \sigma (r_{k+1}^T + \tilde{\tau}_{k+1}^T) - \Gamma \|I - \alpha \sigma \sigma^T\| \hat{W}]\}$$
$$= \frac{1}{\alpha} \text{tr}\{\hat{W}^T \hat{W} + 2\hat{W}^T \alpha \sigma \cdot r_{k+1}^T + 2\hat{W}^T \alpha \sigma \cdot \tilde{\tau}_{k+1}^T - 2\hat{W}^T \Gamma \|I - \alpha \sigma \cdot \sigma^T\| \hat{W}$$
$$+ \alpha^2 r_{k+1} \sigma^T \sigma \cdot r_{k+1}^T + 2\alpha^2 r_{k+1} \sigma^T \sigma \cdot \tilde{\tau}_{k+1}^T - 2\alpha r_{k+1} \sigma^T \Gamma \|I - \alpha \cdot \sigma \cdot \sigma^T\| \hat{W}$$
$$+ \alpha^2 \tilde{\tau}_{k+1} \sigma^T \sigma \cdot \tilde{\tau}_{k+1}^T - 2\alpha \tilde{\tau}_{k+1} \sigma^T \Gamma \|I - \alpha \cdot \sigma \cdot \sigma^T\| \hat{W}$$
$$+ \Gamma^2 \|I - \alpha \cdot \sigma \cdot \sigma^T\|^2 \hat{W}^T \hat{W} + W^T W - \tilde{W}^T \tilde{W}$$
$$- 2W^T \hat{W} - 2W^T \alpha \sigma \cdot r_{k+1}^T - 2W^T \alpha \sigma \cdot \tilde{\tau}_{k+1}^T + 2W^T \Gamma \|I - \alpha \cdot \sigma \cdot \sigma^T\| \hat{W}\}$$
$$= 2r_{k+1}^T \hat{W}^T \sigma + 2\tilde{\tau}_{k+1}^T \hat{W}^T \sigma + \alpha \cdot r_{k+1}^T \sigma^T \sigma r_{k+1} + 2\alpha \tilde{\tau}_{k+1}^T \sigma^T \sigma r_{k+1}$$
$$- 2\Gamma \|I - \alpha \cdot \sigma \sigma^T\| r_{k+1}^T \hat{W}^T \sigma + \alpha \tilde{\tau}_{k+1}^T \sigma^T \sigma \cdot \tilde{\tau}_{k+1} - 2\Gamma \|I - \alpha \cdot \sigma \sigma^T\| \tilde{\tau}_{k+1}^T \hat{W}^T \sigma$$
$$- 2r_{k+1}^T W^T \sigma - 2\tilde{\tau}_{k+1}^T W^T \sigma$$
$$+ \frac{1}{\alpha} \text{tr}\{\hat{W}^T \hat{W} - 2\Gamma \|I - \alpha \cdot \sigma \cdot \sigma^T\| \hat{W}^T \hat{W} + W^T W - \tilde{W}^T \tilde{W}$$
$$- 2W^T \hat{W} + \Gamma^2 \|I - \alpha \cdot \sigma \cdot \sigma^T\|^2 \hat{W}^T \hat{W} + 2W^T \Gamma \|I - \alpha \cdot \sigma \cdot \sigma^T\| \hat{W}\}$$

$$
\begin{aligned}
= \;& 2(r^T K_v^T + D^T - \tilde{\tau}^T)\hat{W}^T\sigma + 2(\tilde{\tau}^T K_b^T + \sigma^T\tilde{W} + \varepsilon^T)\hat{W}^T\sigma \\
& + \alpha \cdot (r^T K_v^T + D^T - \tilde{\tau}^T)\sigma^T\sigma(K_v r + D - \tilde{\tau}) \\
& + 2\alpha(\tilde{\tau}^T K_b^T + \sigma^T\tilde{W} + \varepsilon^T)\sigma^T\sigma(K_v r + D - \tilde{\tau}) \\
& - 2\Gamma\|I - \alpha\cdot\sigma\cdot\sigma^T\|(r^T K_v^T + D^T - \tilde{\tau}^T)\hat{W}^T\sigma + \alpha(\tilde{\tau}^T K_b^T + \sigma^T\tilde{W} + \varepsilon^T)\sigma^T\sigma \\
& \cdot (K_b\tilde{\tau} + \tilde{W}^T\sigma + \varepsilon) - 2\Gamma\|I - \alpha\cdot\sigma\cdot\sigma^T\|(\tilde{\tau}^T K_b^T + \sigma^T\tilde{W} + \varepsilon^T)\hat{W}^T\sigma \\
& - 2(r^T K_v^T + D^T - \tilde{\tau}^T)W^T\sigma - 2(\tilde{\tau}^T K_b^T + \sigma^T\tilde{W} + \varepsilon^T)W^T\sigma \\
& + \frac{1}{\alpha}\,\mathrm{tr}\{-2\Gamma\|I - \alpha\cdot\sigma\cdot\sigma^T\|\hat{W}^T\hat{W} + 2W^T\Gamma\|I - \alpha\cdot\sigma\cdot\sigma^T\|\hat{W} \\
& \qquad + \Gamma^2\|I - \alpha\cdot\sigma\cdot\sigma^T\|^2\hat{W}^T\hat{W}\}.
\end{aligned}
$$

Combining all the terms,

$$
\begin{aligned}
\Delta L = \;& -2r^T[I - K_v^T K_v]r + 4r^T K_v^T D - 4r^T K_v^T\tilde{\tau} + 2D^T D - 4D^T\tilde{\tau} + 2\tilde{\tau}^T\tilde{\tau} \\
& + 2r^T K_v^T K_b\tilde{\tau} + 2r^T K_v^T\tilde{W}^T\sigma + 2r^T K_v^T\varepsilon + 2D^T K_b\tilde{\tau} + 2D^T\tilde{W}^T\sigma + 2D^T\varepsilon \\
& - 2\tilde{\tau}^T K_b^T K_b\tilde{\tau} - 2\tilde{\tau}^T\tilde{W}^T\sigma - 2\tilde{\tau}^T\varepsilon - 2r^T\tilde{\tau} + \tilde{\tau}^T K_b^T K_b\tilde{\tau} + 2\tilde{\tau}^T K_b^T\tilde{W}^T\sigma \\
& + 2\tilde{\tau}^T K_b^T\varepsilon + \sigma^T\tilde{W}\tilde{W}^T\sigma + 2\varepsilon^T\tilde{W}^T\sigma + \varepsilon^T\varepsilon - \tilde{\tau}^T\tilde{\tau} + 2(r^T K_v^T + D^T - \tilde{\tau}^T)\hat{W}^T\sigma \\
& + 2(\tilde{\tau}^T K_b^T + \sigma^T\tilde{W} + \varepsilon^T)\hat{W}^T\sigma + \alpha \cdot (r^T K_v^T + D^T - \tilde{\tau}^T)\sigma^T\sigma(K_v r + D - \tilde{\tau}) \\
& + 2\alpha(\tilde{\tau}^T K_b^T + \sigma^T\tilde{W} + \varepsilon^T)\sigma^T\sigma(K_v r + D - \tilde{\tau}) \\
& - 2\Gamma\|I - \alpha\cdot\sigma\cdot\sigma^T\|(r^T K_v^T + D^T - \tilde{\tau}^T)\hat{W}^T\sigma + \alpha(\tilde{\tau}^T K_b^T + \sigma^T\tilde{W} + \varepsilon^T)\sigma^T\sigma \\
& \cdot (K_b\tilde{\tau} + \tilde{W}^T\sigma + \varepsilon) - 2\Gamma\|I - \alpha\cdot\sigma\cdot\sigma^T\|(\tilde{\tau}^T K_b^T + \sigma^T\tilde{W} + \varepsilon^T)\hat{W}^T\sigma \\
& - 2(r^T K_v^T + D^T - \tilde{\tau}^T)W^T\sigma - 2(\tilde{\tau}^T K_b^T + \sigma^T\tilde{W} + \varepsilon^T)W^T\sigma \\
& + \frac{1}{\alpha}\,\mathrm{tr}\{-2\Gamma\|I - \alpha\cdot\sigma\cdot\sigma^T\|\hat{W}^T\hat{W} + 2W^T\Gamma\|I - \alpha\cdot\sigma\cdot\sigma^T\|\hat{W} \\
& \qquad + \Gamma^2\|I - \alpha\cdot\sigma\cdot\sigma^T\|^2\hat{W}^T\hat{W}\}.
\end{aligned}
$$

Simplifying,

$$
\begin{aligned}
\Delta L = \;& -2r^T[I - K_v^T K_v]r + 4r^T K_v^T D - 4r^T K_v^T\tilde{\tau} + 2D^T D - 4D^T\tilde{\tau} + \tilde{\tau}^T\tilde{\tau} \\
& + 2r^T K_v^T K_b\tilde{\tau} + 2r^T K_v^T\tilde{W}^T\sigma + 2r^T K_v^T\varepsilon + 2D^T K_b\tilde{\tau} + 2D^T\tilde{W}^T\sigma + 2D^T\varepsilon \\
& - \tilde{\tau}^T K_b^T K_b\tilde{\tau} - 2\tilde{\tau}^T\tilde{W}^T\sigma - 2\tilde{\tau}^T\varepsilon - 2r^T\tilde{\tau} + 2\tilde{\tau}^T K_b^T\tilde{W}^T\sigma + 2\tilde{\tau}^T K_b^T\varepsilon \\
& + \sigma^T\tilde{W}\tilde{W}^T\sigma + 2\varepsilon^T\tilde{W}^T\sigma + \varepsilon^T\varepsilon - 2(r^T K_v^T + D^T - \tilde{\tau}^T)\tilde{W}^T\sigma \\
& - 2(\tilde{\tau}^T K_b^T + \sigma^T\tilde{W} + \varepsilon^T)\tilde{W}^T\sigma + \alpha \cdot (r^T K_v^T + D^T - \tilde{\tau}^T)\sigma^T\sigma(K_v r + D - \tilde{\tau}) \\
& + 2\alpha(\tilde{\tau}^T K_b^T + \sigma^T\tilde{W} + \varepsilon^T)\sigma^T\sigma(K_v r + D - \tilde{\tau}) \\
& - 2\Gamma\|I - \alpha\sigma\sigma^T\|(r^T K_v^T + D^T - \tilde{\tau}^T)(W - \tilde{W})^T\sigma \\
& + \alpha(\tilde{\tau}^T K_b^T + \sigma^T\tilde{W} + \varepsilon^T)\sigma^T\sigma(K_b\tilde{\tau} + \tilde{W}^T\sigma + \varepsilon) \\
& - 2\Gamma\|I - \alpha\cdot\sigma\sigma^T\|(\tilde{\tau}^T K_b^T + \sigma^T\tilde{W} + \varepsilon^T)(W - \tilde{W})^T\sigma \\
& + \frac{1}{\alpha}\,\mathrm{tr}\{-2\Gamma\|I - \alpha\cdot\sigma\sigma^T\|\hat{W}^T\hat{W} + 2W^T\Gamma\|I - \alpha\cdot\sigma\sigma^T\|\hat{W} \\
& \qquad + \Gamma^2\|I - \alpha\cdot\sigma\sigma^T\|^2\hat{W}^T\hat{W}\}.
\end{aligned}
$$

Multiplying out terms,

$$\Delta L = -2r^T[I - K_v^T K_v]r + 4r^T K_v^T D - 4r^T K_v^T \tilde{\tau} + 2D^T D - 4D^T \tilde{\tau} + \tilde{\tau}^T \tilde{\tau}$$
$$+ 2r^T K_v^T K_b \tilde{\tau} + 2r^T K_v^T \tilde{W}^T \sigma + 2r^T K_v^T \varepsilon + 2D^T K_b \tilde{\tau} + 2D^T \tilde{W}^T \sigma + 2D^T \varepsilon$$
$$- \tilde{\tau}^T K_b^T K_b \tilde{\tau} - 2\tilde{\tau}^T \tilde{W}^T \sigma - 2\tilde{\tau}^T \varepsilon - 2r^T \tilde{\tau} + 2\tilde{\tau}^T K_b^T \tilde{W}^T \sigma + 2\tilde{\tau}^T K_b^T \varepsilon + \sigma^T \tilde{W} \tilde{W}^T \sigma$$
$$+ 2\varepsilon^T \tilde{W}^T \sigma + \varepsilon^T \varepsilon - 2r^T K_v^T \tilde{W}^T \sigma - 2D^T \tilde{W}^T \sigma + 2\tilde{\tau}^T \tilde{W}^T \sigma - 2\tilde{\tau}^T K_b^T \tilde{W}^T \sigma$$
$$- 2\sigma^T \tilde{W} \tilde{W}^T \sigma - 2\varepsilon^T \tilde{W}^T \sigma + \alpha r^T K_v^T \sigma^T \sigma K_v r + 2\alpha r^T K_v^T \sigma^T \sigma D - 2\alpha r^T K_v^T \sigma^T \sigma \tilde{\tau}$$
$$+ \alpha D^T \sigma^T \sigma D - 2\alpha D^T \sigma^T \sigma \tilde{\tau} + \alpha \tilde{\tau}^T \sigma^T \sigma \tilde{\tau} + 2\alpha \tilde{\tau}^T K_b^T \sigma^T \sigma K_v r + 2\alpha \tilde{\tau}^T K_b^T \sigma^T \sigma D$$
$$- 2\alpha \tilde{\tau}^T K_b^T \sigma^T \sigma \tilde{\tau} + 2\alpha \sigma^T \tilde{W} \sigma^T \sigma K_v r + 2\alpha \sigma^T \tilde{W} \sigma^T \sigma D - 2\alpha \sigma^T \tilde{W} \sigma^T \sigma \tilde{\tau}$$
$$+ 2\alpha \varepsilon^T \sigma^T \sigma K_v r + 2\alpha \varepsilon^T \sigma^T \sigma D - 2\alpha \varepsilon^T \sigma^T \sigma \tilde{\tau} - 2\Gamma \|I - \alpha \cdot \sigma \sigma^T\| r^T K_v^T W^T \sigma$$
$$+ 2\Gamma \|I - \alpha \cdot \sigma \sigma^T\| r^T K_v^T \tilde{W}^T \sigma - 2\Gamma \|I - \alpha \cdot \sigma \sigma^T\| D^T W^T \sigma$$
$$+ 2\Gamma \|I - \alpha \cdot \sigma \sigma^T\| D^T \tilde{W}^T \sigma + 2\Gamma \|I - \alpha \cdot \sigma \sigma^T\| \tilde{\tau}^T W^T \sigma$$
$$- 2\Gamma \|I - \alpha \cdot \sigma \sigma^T\| \tilde{\tau}^T \tilde{W}^T \sigma + \alpha \tilde{\tau}^T K_b^T \sigma^T \sigma K_b \tilde{\tau} + 2\alpha \tilde{\tau}^T K_b^T \sigma^T \sigma \tilde{W}^T \sigma$$
$$+ 2\alpha \tilde{\tau}^T K_b^T \sigma^T \sigma \varepsilon + \alpha \sigma^T \tilde{W} \sigma^T \sigma \tilde{W}^T \sigma + 2\alpha \sigma^T \tilde{W} \sigma^T \sigma \varepsilon + \alpha \varepsilon^T \sigma^T \sigma \cdot \varepsilon$$
$$- 2\Gamma \|I - \alpha \cdot \sigma \cdot \sigma^T\| \tilde{\tau}^T K_b^T W^T \sigma + 2\Gamma \|I - \alpha \cdot \sigma \sigma^T\| \tilde{\tau}^T K_b^T \tilde{W}^T \sigma$$
$$- 2\Gamma \|I - \alpha \cdot \sigma \sigma^T\| \sigma^T \tilde{W} \hat{W}^T \sigma - 2\Gamma \|I - \alpha \cdot \sigma \sigma^T\| \varepsilon^T W^T \sigma$$
$$+ 2\Gamma \|I - \alpha \cdot \sigma \sigma^T\| \varepsilon^T \tilde{W}^T \sigma$$
$$+ \frac{1}{\alpha} \text{tr}\{-2\Gamma \|I - \alpha \cdot \sigma \sigma^T\| \hat{W}^T \hat{W} + 2W^T \Gamma \|I - \alpha \cdot \sigma \sigma^T\| \hat{W}$$
$$+ \Gamma^2 \|I - \alpha \cdot \sigma \sigma^T\|^2 \hat{W}^T \hat{W}\}.$$

Simplifying,

$$\Delta L = -r^T[2I - (2 + \alpha \sigma^T \sigma)K_v^T K_v]r + 2(2 + \alpha \sigma^T \sigma)r^T K_v^T D + (2 + \alpha \sigma^T \sigma)D^T D$$
$$+ 2(1 + \alpha \sigma^T \sigma)r^T K_v^T \varepsilon + (1 + \alpha \sigma^T \sigma)\varepsilon^T \varepsilon + 2(1 + \alpha \sigma^T \sigma)\varepsilon^T D + \tilde{\tau}^T \tilde{\tau}$$
$$- \tilde{\tau}^T K_b^T K_b \tilde{\tau} + \alpha \tilde{\tau}^T K_b^T \sigma^T \sigma K_b \tilde{\tau} + \alpha \tilde{\tau}^T \sigma^T \sigma \tilde{\tau} - 2\alpha \tilde{\tau}^T K_b^T \sigma^T \sigma \tilde{\tau}$$
$$+ 2(1 + \alpha \sigma^T \sigma)r^T K_v^T K_b \tilde{\tau} - 2r^T \tilde{\tau} + 2(1 + \alpha \sigma^T \sigma)D^T K_b \tilde{\tau} - 2(1 + \alpha \sigma^T \sigma)\tilde{\tau}^T \varepsilon$$
$$- 2(2 + \alpha \sigma^T \sigma)r^T K_v^T \tilde{\tau} - 2\alpha \sigma^T \tilde{W} \sigma^T \sigma \tilde{\tau} - 2(2 + \alpha \sigma^T \sigma)D^T \tilde{\tau}$$
$$+ 2\alpha \tilde{\tau}^T K_b^T \sigma^T \sigma \tilde{W}^T \sigma + 2(1 + \alpha \sigma^T \sigma)\tilde{\tau}^T K_b^T \varepsilon + 2\alpha \sigma^T \sigma r^T K_v^T \tilde{W}^T \sigma$$
$$- (1 - \alpha \sigma^T \sigma)\sigma^T \tilde{W} \tilde{W}^T \sigma + 2\alpha \sigma^T \sigma \varepsilon^T \tilde{W}^T \sigma + 2\alpha \sigma^T \tilde{W} \sigma^T \sigma D$$
$$+ 2\Gamma \|I - \alpha \cdot \sigma \sigma^T\| D^T \tilde{W}^T \sigma + 2\Gamma \|I - \alpha \cdot \sigma \sigma^T\| r^T K_v^T \tilde{W}^T \sigma$$
$$- 2\Gamma \|I - \alpha \cdot \sigma \sigma^T\| r^T K_v^T W^T \sigma - 2\Gamma \|I - \alpha \cdot \sigma \sigma^T\| D^T W^T \sigma$$
$$+ 2\Gamma \|I - \alpha \cdot \sigma \sigma^T\| \varepsilon^T \tilde{W}^T \sigma - 2\Gamma \|I - \alpha \cdot \sigma \sigma^T\| \varepsilon^T W^T \sigma$$
$$- 2\Gamma \|I - \alpha \cdot \sigma \sigma^T\| \tilde{\tau}^T K_b^T W^T \sigma + 2\Gamma \|I - \alpha \cdot \sigma \sigma^T\| \tilde{\tau}^T K_b^T \tilde{W}^T \sigma$$
$$+ 2\Gamma \|I - \alpha \cdot \sigma \sigma^T\| \tilde{\tau}^T W^T \sigma - 2\Gamma \|I - \alpha \cdot \sigma \sigma^T\| \tilde{\tau}^T \tilde{W}^T \sigma$$
$$- 2\Gamma \|I - \alpha \cdot \sigma \sigma^T\| \sigma^T \tilde{W} \hat{W}^T \sigma$$
$$+ \frac{1}{\alpha} \text{tr}\{-2\Gamma \|I - \alpha \cdot \sigma \sigma^T\| \hat{W}^T \hat{W} + 2W^T \Gamma \|I - \alpha \cdot \sigma \sigma^T\| \hat{W}$$
$$+ \Gamma^2 \|I - \alpha \cdot \sigma \sigma^T\|^2 \hat{W}^T \hat{W}\}.$$

Pick $K_b = (I + K_v^{-1})^T = I + K_a$ and define $\beta = 2K_a + (1 - \alpha\sigma^T\sigma)K_a^T K_a - I > 0$ (condition (4.2.32)), which is true as long as $K_v^{-1} < I$ (condition (4.2.31)). It can be seen that $\beta > I$ and β is a diagonal matrix since K_a is diagonal:

$$
\begin{aligned}
\Delta L = &-r^T[2I - (2 + \alpha\sigma^T\sigma)K_v^T K_v]r + 2(2 + \alpha\sigma^T\sigma)r^T K_v^T D + (2 + \alpha\sigma^T\sigma)D^T D \\
&+ 2(1 + \alpha\sigma^T\sigma)r^T K_v^T \varepsilon + (1 + \alpha\sigma^T\sigma)\varepsilon^T \varepsilon + 2(1 + \alpha\sigma^T\sigma)\varepsilon^T D - (\beta + I)\tilde{\tau}^T\tilde{\tau} \\
&+ 2(1 + \alpha\sigma^T\sigma)r^T K_v^T(I + K_a)\tilde{\tau} - 2r^T\tilde{\tau} + 2(1 + \alpha\sigma^T\sigma)D^T(I + K_a)\tilde{\tau} \\
&- 2(1 + \alpha\sigma^T\sigma)\tilde{\tau}^T\varepsilon - 2(2 + \alpha\sigma^T\sigma)r^T K_v^T\tilde{\tau} - 2\alpha\sigma^T\tilde{W}\sigma^T\sigma\tilde{\tau} - 2(2 + \alpha\sigma^T\sigma)D^T\tilde{\tau} \\
&+ 2\alpha\tilde{\tau}^T(I + K_a^T)\sigma^T\sigma\tilde{W}^T\sigma + 2(1 + \alpha\sigma^T\sigma)\tilde{\tau}^T(I + K_a^T)\varepsilon + 2\alpha\sigma^T\sigma r^T K_v^T\tilde{W}^T\sigma \\
&- (1 - \alpha\sigma^T\sigma)\sigma^T\tilde{W}\tilde{W}^T\sigma + 2\alpha\sigma^T\sigma\varepsilon^T\tilde{W}^T\sigma + 2\alpha\sigma^T\tilde{W}\sigma^T\sigma D \\
&+ 2\Gamma\|I - \alpha\cdot\sigma\sigma^T\|D^T\tilde{W}^T\sigma + 2\Gamma\|I - \alpha\cdot\sigma\sigma^T\|r^T K_v^T\tilde{W}^T\sigma \\
&- 2\Gamma\|I - \alpha\cdot\sigma\sigma^T\|r^T K_v^T W^T\sigma - 2\Gamma\|I - \alpha\cdot\sigma\sigma^T\|D^T W^T\sigma \\
&+ 2\Gamma\|I - \alpha\cdot\sigma\cdot\sigma^T\|\varepsilon^T\tilde{W}^T\sigma - 2\Gamma\|I - \alpha\cdot\sigma\sigma^T\|\varepsilon^T W^T\sigma \\
&- 2\Gamma\|I - \alpha\cdot\sigma\sigma^T\|\sigma^T\tilde{W}\hat{W}^T\sigma - 2\Gamma\|I - \alpha\cdot\sigma\sigma^T\|\tilde{\tau}^T(I + K_a^T)W^T\sigma \\
&+ 2\Gamma\|I - \alpha\cdot\sigma\sigma^T\|\tau^T(I + K_a^T)\tilde{W}^T\sigma + 2\Gamma\|I - \alpha\cdot\sigma\sigma^T\|\tau^T W^T\sigma \\
&- 2\Gamma\|I - \alpha\cdot\sigma\sigma^T\|\tilde{\tau}^T\tilde{W}^T\sigma \\
&+ \frac{1}{\alpha}\mathrm{tr}\{-2\Gamma\|I - \alpha\cdot\sigma\sigma^T\|\hat{W}^T\hat{W} + 2W^T\Gamma\|I - \alpha\cdot\sigma\sigma^T\|\hat{W} \\
&\qquad + \Gamma^2\|I - \alpha\cdot\sigma\sigma^T\|^2\hat{W}^T\hat{W}\},
\end{aligned}
$$

$$
\begin{aligned}
\Delta L = &-r^T[2I - (2 + \alpha\sigma^T\sigma)K_v^T K_v]r + 2(2 + \alpha\sigma^T\sigma)r^T K_v^T D + (2 + \alpha\sigma^T\sigma)D^T D \\
&+ 2(1 + \alpha\sigma^T\sigma)r^T K_v^T\varepsilon + (1 + \alpha\sigma^T\sigma)\varepsilon^T\varepsilon + 2(1 + \alpha\sigma^T\sigma)\varepsilon^T D - (I + \beta)\tilde{\tau}^T\tau \\
&- 2r^T K_v^T\tilde{\tau} + 2\alpha\sigma^T\sigma r^T\tilde{\tau} + 2(1 + \alpha\sigma^T\sigma)D^T K_a\tilde{\tau} - 2D^T\tilde{\tau} \\
&+ 2\alpha\tilde{\tau}^T K_a^T\sigma^T\sigma\tilde{W}^T\sigma + 2(1 + \alpha\sigma^T\sigma)\tilde{\tau}^T K_a - 2\Gamma\|I - \alpha\cdot\sigma\sigma^T\|\tilde{\tau}^T K_a^T W^T\sigma \\
&+ 2\Gamma\|I - \alpha\cdot\sigma\sigma^T\|\tilde{\tau}^T K_a^T\tilde{W}^T\sigma + 2\alpha\sigma^T\sigma r^T K_v^T\tilde{W}^T\sigma - (1 - \alpha\sigma^T\sigma)\sigma^T\tilde{W}\tilde{W}^T\sigma \\
&+ 2\alpha\sigma^T\sigma\varepsilon^T\tilde{W}^T\sigma + 2\alpha\sigma^T\tilde{W}\sigma^T\sigma D + 2\Gamma\|I - \alpha\cdot\sigma\sigma^T\|D^T\tilde{W}^T\sigma \\
&+ 2\Gamma\|I - \alpha\cdot\sigma\sigma^T\|r^T K_v^T\tilde{W}^T\sigma - 2\Gamma\|I - \alpha\cdot\sigma\sigma^T\|r^T K_v^T W^T\sigma \\
&- 2\Gamma\|I - \alpha\cdot\sigma\sigma^T\|D^T W^T\sigma + 2\Gamma\|I - \alpha\cdot\sigma\sigma^T\|\varepsilon^T\tilde{W}^T\sigma \\
&- 2\Gamma\|I - \alpha\cdot\sigma\sigma^T\|\varepsilon^T W^T\sigma - 2\Gamma\|I - \alpha\cdot\sigma\sigma^T\|\sigma^T\tilde{W}\hat{W}^T\sigma \\
&+ \frac{1}{\alpha}\mathrm{tr}\{-2\Gamma\|I - \alpha\cdot\sigma\sigma^T\|\hat{W}^T\hat{W} + 2W^T\Gamma\|I - \alpha\cdot\sigma\sigma^T\|\hat{W} \\
&\qquad + \Gamma^2\|I - \alpha\cdot\sigma\sigma^T\|^2\hat{W}^T\hat{W}\}.
\end{aligned}
$$

Completing squares for $K_a\tilde{\tau}$ and $\tilde{\tau}$,

$$
\begin{aligned}
\Delta L = &-r^T[2I - (2 + \alpha\sigma^T\sigma)K_v^T K_v]r + 2(2 + \alpha\sigma^T\sigma)r^T K_v^T D + (2 + \alpha\sigma^T\sigma)D^T D \\
&+ 2(1 + \alpha\sigma^T\sigma)r^T K_v^T\varepsilon + (1 + \alpha\sigma^T\sigma)\varepsilon^T\varepsilon + 2(1 + \alpha\sigma^T\sigma)\varepsilon^T D \\
&- \{K_a\tilde{\tau} - \beta^{-1}[(1 + \alpha\sigma^T\sigma)(D + \varepsilon) - (\alpha\sigma^T\sigma + \Gamma\|I - \alpha\sigma\sigma^T\|)\tilde{W}^T\sigma \\
&\quad + \Gamma\|I - \alpha\sigma\sigma^T\|W^T\sigma]\}^T \\
&\cdot \beta\cdot\{K_a\tilde{\tau} - \beta^{-1}[(1 + \alpha\sigma^T\sigma)(D + \varepsilon) - (\alpha\sigma^T\sigma + \Gamma\|I - \alpha\sigma\sigma^T\|)\tilde{W}^T\sigma \\
&\quad + \Gamma\|I - \alpha\sigma\sigma^T\|W^T\sigma]\}
\end{aligned}
$$

$$+ \{\beta^{-1}[(1 + \alpha\sigma^T\sigma)(D + \varepsilon) + (\alpha\sigma^T\sigma + \Gamma\|I - \alpha\sigma\sigma^T\|)\tilde{W}^T\sigma$$
$$- \Gamma\|I - \alpha\sigma\sigma^T\|W^T\sigma]\}^T$$
$$\cdot \beta \cdot \{\beta^{-1}[(1 + \alpha\sigma^T\sigma)(D + \varepsilon) + (\alpha\sigma^T\sigma + \Gamma\|I - \alpha\sigma\sigma^T\|)\tilde{W}^T\sigma$$
$$- \Gamma\|I - \alpha\sigma\sigma^T\|W^T\sigma]\}$$
$$- [\tilde{\tau} - \alpha\sigma^T\sigma r + D + K_v r]^T[\tilde{\tau} - \alpha\sigma^T\sigma r + D + K_v r] + (\alpha\sigma^T\sigma)^2 r^T r$$
$$+ r^T K_v^T K_v r + D^T D - 2\alpha\sigma^T\sigma r^T D + 2r^T K_v^T D - 2\alpha\sigma^T\sigma r^T K_v r$$
$$+ 2\alpha\sigma^T\sigma r^T K_v^T \tilde{W}^T\sigma - (1 - \alpha\sigma^T\sigma)\sigma^T \tilde{W}\tilde{W}^T\sigma + 2\alpha\sigma^T\sigma\varepsilon^T\tilde{W}^T\sigma$$
$$+ 2\alpha\sigma^T\tilde{W}\sigma^T\sigma D + 2\Gamma\|I - \alpha\cdot\sigma\cdot\sigma^T\|D^T\tilde{W}^T\sigma$$
$$+ 2\Gamma\|I - \alpha\cdot\sigma\cdot\sigma^T\|r^T K_v^T \tilde{W}^T\sigma - 2\Gamma\|I - \alpha\cdot\sigma\sigma^T\|r^T K_v^T W^T\sigma$$
$$- 2\Gamma\|I - \alpha\cdot\sigma\sigma^T\|D^T W^T\sigma + 2\Gamma\|I - \alpha\cdot\sigma\sigma^T\|\varepsilon^T\tilde{W}^T\sigma$$
$$- 2\Gamma\|I - \alpha\cdot\sigma\sigma^T\|\varepsilon^T W^T\sigma - 2\Gamma\|I - \alpha\cdot\sigma\sigma^T\|\sigma^T\tilde{W}\hat{W}^T\sigma$$
$$+ \frac{1}{\alpha}\text{tr}\{-2\Gamma\|I - \alpha\cdot\sigma\sigma^T\|\hat{W}^T\hat{W} + 2W^T\Gamma\|I - \alpha\cdot\sigma\sigma^T\|\hat{W}$$
$$+ \Gamma^2\|I - \alpha\cdot\sigma\sigma^T\|^2\hat{W}^T\hat{W}\},$$
$$\Delta L = -r^T[2I - (3 + \alpha\sigma^T\sigma)K_v^T K_v - (\alpha\sigma^T\sigma)^2 + 2\alpha\sigma^T\sigma K_v]r + 2(3 + \alpha\sigma^T\sigma)r^T K_v^T D$$
$$+ (3 + \alpha\sigma^T\sigma)D^T D + 2(1 + \alpha\sigma^T\sigma)r^T K_v^T\varepsilon + (1 + \alpha\sigma^T\sigma)\varepsilon^T\varepsilon$$
$$+ 2(1 + \alpha\sigma^T\sigma)\varepsilon^T D$$
$$- \{K_a\tilde{\tau} - \beta^{-1}[(1 + \alpha\sigma^T\sigma)(D + \varepsilon) - (\alpha\sigma^T\sigma + \Gamma\|I - \alpha\sigma\sigma^T\|)\tilde{W}^T\sigma$$
$$+ \Gamma\|I - \alpha\sigma\sigma^T\|W^T\sigma]\}^T$$
$$\cdot \beta \cdot \{K_a\tilde{\tau} - \beta^{-1}[(1 + \alpha\sigma^T\sigma)(D + \varepsilon) - (\alpha\sigma^T\sigma + \Gamma\|I - \alpha\sigma\sigma^T\|)\tilde{W}^T\sigma$$
$$+ \Gamma\|I - \alpha\sigma\sigma^T\|W^T\sigma]\}$$
$$- [\tilde{\tau} - \alpha\sigma^T\sigma r + D + K_v r]^T[\tilde{\tau} - \alpha\sigma^T\sigma r + D + K_v r]$$
$$+ \beta^{-1}(1 + \alpha\sigma^T\sigma)^2(D + \varepsilon)^T(D + \varepsilon) - 2\alpha\sigma^T\sigma r^T D$$
$$+ 2\beta^{-1}(\alpha\sigma^T\sigma + \Gamma\|I - \alpha\sigma\sigma^T\|)(1 + \alpha\sigma^T\sigma)(D + \varepsilon)^T\tilde{W}^T\sigma$$
$$+ \beta^{-1}(\alpha\sigma^T\sigma + \Gamma\|I - \alpha\sigma\sigma^T\|)^2\sigma^T\tilde{W}\tilde{W}^T\sigma + \beta^{-1}\Gamma^2\|I - \alpha\sigma\sigma^T\|^2\sigma^T W W^T\sigma$$
$$- 2\beta^{-1}(1 + \alpha\sigma^T\sigma)\Gamma\|I - \alpha\sigma\sigma^T\|(D + \varepsilon)^T W^T\sigma$$
$$- 2\beta^{-1}(\alpha\sigma^T\sigma + \Gamma\|I - \alpha\sigma\sigma^T\|)\Gamma\|I - \alpha\sigma\sigma^T\|\sigma^T\tilde{W}W^T\sigma$$
$$+ 2\alpha\sigma^T\sigma r^T K_v^T \tilde{W}^T\sigma - (1 - \alpha\sigma^T\sigma)\sigma^T\tilde{W}\tilde{W}^T\sigma + 2\alpha\sigma^T\sigma\varepsilon^T\tilde{W}^T\sigma$$
$$+ 2\alpha\sigma^T\tilde{W}\sigma^T\sigma D + 2\Gamma\|I - \alpha\cdot\sigma\sigma^T\|D^T\tilde{W}^T\sigma + 2\Gamma\|I - \alpha\cdot\sigma\sigma^T\|r^T K_v^T\tilde{W}^T\sigma$$
$$- 2\Gamma\|I - \alpha\cdot\sigma\sigma^T\|r^T K_v^T W^T\sigma - 2\Gamma\|I - \alpha\cdot\sigma\cdot\sigma^T\|D^T W^T\sigma$$
$$+ 2\Gamma\|I - \alpha\cdot\sigma\sigma^T\|\varepsilon^T\tilde{W}^T\sigma - 2\Gamma\|I - \alpha\cdot\sigma\sigma^T\|\varepsilon^T W^T\sigma$$
$$- 2\Gamma\|I - \alpha\cdot\sigma\sigma^T\|\sigma^T\tilde{W}\hat{W}^T\sigma$$
$$+ \frac{1}{\alpha}\text{tr}\{-2\Gamma\|I - \alpha\cdot\sigma\sigma^T\|\hat{W}^T\hat{W} + 2W^T\Gamma\|I - \alpha\cdot\sigma\sigma^T\|\hat{W}$$
$$+ \Gamma^2\|I - \alpha\cdot\sigma\sigma^T\|^2\hat{W}^T\hat{W}\},$$
$$\Delta L = -r^T[2I - (3 + \alpha\sigma^T\sigma)K_v^T K_v - (\alpha\sigma^T\sigma)^2 + 2\alpha\sigma^T\sigma K_v]r$$
$$+ 2(1 + \alpha\sigma^T\sigma)r^T K_v^T(D + \varepsilon) + 4r^T K_v^T D + 2D^T D$$

$$
\begin{aligned}
&- \{K_a \tilde{\tau} - \beta^{-1}[(1 + \alpha \sigma^T \sigma)(D + \varepsilon) - (\alpha \sigma^T \sigma + \Gamma \| I - \alpha \sigma \sigma^T \|) \tilde{W}^T \sigma \\
&\qquad\qquad + \Gamma \| I - \alpha \sigma \sigma^T \| W^T \sigma] \}^T \\
&\cdot \beta \{ K_a \tilde{\tau} - \beta^{-1}[(1 + \alpha \sigma^T \sigma)(D + \varepsilon) - (\alpha \sigma^T \sigma + \Gamma \| I - \alpha \sigma \sigma^T \|) \tilde{W}^T \sigma \\
&\qquad\qquad + \Gamma \| I - \alpha \sigma \sigma^T \| W^T \sigma] \} - 2 \alpha \sigma^T \sigma r^T D \\
&- [\tilde{\tau} - \alpha \sigma^T \sigma r + D + K_v r]^T [\tilde{\tau} - \alpha \sigma^T \sigma r + D + K_v r] \\
&+ (1 + \alpha \sigma^T \sigma)[1 + \beta^{-1}(1 + \alpha \sigma^T \sigma)](D + \varepsilon)^T (D + \varepsilon) \\
&+ 2(\alpha \sigma^T \sigma + \Gamma \| I - \alpha \cdot \sigma \sigma^T \|) r^T K_v^T \tilde{W}^T \sigma \\
&- [1 - \alpha \sigma^T \sigma - \beta^{-1}(\alpha \sigma^T \sigma + \Gamma \| I - \alpha \sigma \sigma^T \|)^2] \sigma^T \tilde{W} \tilde{W}^T \sigma \\
&+ \beta^{-1} \Gamma^2 \| I - \alpha \sigma \sigma^T \|^2 \sigma^T W W^T \sigma \\
&- 2 \beta^{-1}(1 + \alpha \sigma^T \sigma) \Gamma \| I - \alpha \sigma \sigma^T \| (D + \varepsilon)^T W^T \sigma \\
&- 2 \beta^{-1}(\alpha \sigma^T \sigma + \Gamma \| I - \alpha \sigma \sigma^T \|) \Gamma \| I - \alpha \sigma \sigma^T \| \sigma^T \tilde{W} W^T \sigma \\
&+ 2(\alpha \sigma^T \sigma + \Gamma \| I - \alpha \sigma \sigma^T \|)[1 + \beta^{-1}(1 + \alpha \sigma^T \sigma)](D + \varepsilon)^T \tilde{W}^T \sigma \\
&- 2 \Gamma \| I - \alpha \cdot \sigma \sigma^T \| r^T K_v^T W^T \sigma - 2 \Gamma \| I - \alpha \cdot \sigma \sigma^T \| (D + \varepsilon)^T W^T \sigma \\
&- 2 \Gamma \| I - \alpha \cdot \sigma \sigma^T \| \sigma^T \tilde{W} \hat{W}^T \sigma \\
&+ \frac{1}{\alpha} \operatorname{tr} \{ -2 \Gamma \| I - \alpha \cdot \sigma \sigma^T \| \hat{W}^T \hat{W} + 2 W^T \Gamma \| I - \alpha \cdot \sigma \sigma^T \| \hat{W} \\
&\qquad\qquad + \Gamma^2 \| I - \alpha \cdot \sigma \sigma^T \|^2 \hat{W}^T \hat{W} \}.
\end{aligned}
$$

Define $\rho = (1 - \alpha \sigma^T \sigma) I - \beta^{-1}(\alpha \sigma^T \sigma + \Gamma \| I - \alpha \sigma \sigma^T \|)^2 > 0$ (condition (5.3.30)). Completing squares for $\tilde{W}^T \sigma$,

$$
\begin{aligned}
\Delta L = &- r^T [2I - (3 + \alpha \sigma^T \sigma) K_v^T K_v - (\alpha \sigma^T \sigma)^2 + 2 \alpha \sigma^T \sigma K_v] r \\
&+ 2(1 + \alpha \sigma^T \sigma) r^T K_v^T (D + \varepsilon) + 4 r^T K_v^T D + 2 D^T D \\
&- \{ K_a \tilde{\tau} - \beta^{-1}[(1 + \alpha \sigma^T \sigma)(D + \varepsilon) - (\alpha \sigma^T \sigma + \Gamma \| I - \alpha \sigma \sigma^T \|) \tilde{W}^T \sigma \\
&\qquad\qquad + \Gamma \| I - \alpha \sigma \sigma^T \| W^T \sigma] \}^T \cdot \beta \\
&\cdot \{ K_a \tilde{\tau} - \beta^{-1}[(1 + \alpha \sigma^T \sigma)(D + \varepsilon) - (\alpha \sigma^T \sigma + \Gamma \| I - \alpha \sigma \sigma^T \|) \tilde{W}^T \sigma \\
&\qquad\qquad + \Gamma \| I - \alpha \sigma \sigma^T \| W^T \sigma] \} \\
&- [\tilde{\tau} - \alpha \sigma^T \sigma r + D + K_v r]^T [\tilde{\tau} - \alpha \sigma^T \sigma r + D + K_v r] - 2 \alpha \sigma^T \sigma r^T D \\
&+ (1 + \alpha \sigma^T \sigma) \left(1 + \frac{1 + \alpha \sigma^T \sigma}{\beta} \right) (D + \varepsilon)^T (D + \varepsilon) \\
&- \{ \tilde{W}^T \sigma - \rho^{-1}(\alpha \sigma^T \sigma + \Gamma \| I - \alpha \sigma \sigma^T \|) [K_v r + (1 + \beta^{-1}(1 + \alpha \sigma^T \sigma))(D + \varepsilon) \\
&\qquad\qquad\qquad\qquad - \beta^{-1} \Gamma \| I - \alpha \sigma \sigma^T \| W^T \sigma] \}^T \rho \\
&\cdot \{ \tilde{W}^T \sigma - \rho^{-1}(\alpha \sigma^T \sigma + \Gamma \| I - \alpha \sigma \sigma^T \|) [K_v r + (1 + \beta^{-1}(1 + \alpha \sigma^T \sigma))(D + \varepsilon) \\
&\qquad\qquad\qquad\qquad - \beta^{-1} \Gamma \| I - \alpha \sigma \sigma^T \| W^T \sigma] \} \\
&+ \rho^{-1}(\alpha \sigma^T \sigma + \Gamma \| I - \alpha \sigma \sigma^T \|)^2 \\
&\cdot [K_v r + (1 + \beta^{-1}(1 + \alpha \sigma^T \sigma))(D + \varepsilon) - \beta^{-1} \Gamma \| I - \alpha \sigma \sigma^T \| W^T \sigma]^T \\
&\cdot [K_v r + (1 + \beta^{-1}(1 + \alpha \sigma^T \sigma))(D + \varepsilon) - \beta^{-1} \Gamma \| I - \alpha \sigma \sigma^T \| W^T \sigma] \\
&+ \beta^{-1} \Gamma^2 \| I - \alpha \sigma \sigma^T \|^2 \sigma^T W W^T \sigma \\
&- 2 \beta^{-1}(1 + \alpha \sigma^T \sigma) \Gamma \| I - \alpha \sigma \sigma^T \| (D + \varepsilon)^T W^T \sigma
\end{aligned}
$$

$$-2\Gamma\|I-\alpha\cdot\sigma\cdot\sigma^T\|r^T K_v^T W^T\sigma - 2\Gamma\|I-\alpha\cdot\sigma\cdot\sigma^T\|(D+\varepsilon)^T W^T\sigma$$

$$-2\Gamma\|I-\alpha\cdot\sigma\cdot\sigma^T\|\sigma^T\tilde{W}\hat{W}^T\sigma$$

$$+\frac{1}{\alpha}\mathrm{tr}\{-2\Gamma\|I-\alpha\cdot\sigma\cdot\sigma^T\|\hat{W}^T\hat{W} + 2W^T\Gamma\|I-\alpha\cdot\sigma\cdot\sigma^T\|\hat{W}$$

$$+\Gamma^2\|I-\alpha\cdot\sigma\cdot\sigma^T\|^2\hat{W}^T\hat{W}\}.$$

Putting the term $-2\Gamma\|I-\alpha\cdot\sigma^T\sigma\|\sigma^T\tilde{W}\hat{W}^T\sigma$ back on the trace term and bounding the trace term,

$$\frac{1}{\alpha}\mathrm{tr}\{-2\alpha\Gamma\|I-\alpha\cdot\sigma\sigma^T\|\tilde{W}^T\sigma\sigma^T\hat{W} + 2\Gamma\|I-\alpha\cdot\sigma\sigma^T\|\tilde{W}^T\hat{W}$$

$$+\|I-\alpha\cdot\sigma\sigma^T\|^2\hat{W}^T\Gamma^T\Gamma\hat{W}\}$$

$$=-\frac{\|I-\alpha\cdot\sigma\sigma^T\|}{\alpha}\mathrm{tr}\{2\alpha\Gamma\tilde{W}^T\sigma\sigma^T\hat{W} - 2\Gamma\tilde{W}^T\hat{W} - \|I-\alpha\cdot\sigma\sigma^T\|\hat{W}^T\Gamma^T\Gamma\hat{W}\}$$

$$=-\frac{\|I-\alpha\cdot\sigma\sigma^T\|}{\alpha}\mathrm{tr}\{2\alpha\Gamma\tilde{W}^T\sigma\sigma^T(W-\hat{W}) - 2\Gamma\tilde{W}^T(W-\hat{W})$$

$$-\|I-\alpha\cdot\sigma\sigma^T\|(W-\hat{W})^T\Gamma^T\Gamma(W-\hat{W})\}$$

$$=-\frac{\|I-\alpha\cdot\sigma\sigma^T\|}{\alpha}\mathrm{tr}\{2\alpha\Gamma\tilde{W}^T\sigma\sigma^T W - 2\alpha\Gamma\tilde{W}^T\sigma\sigma^T\tilde{W} - 2\Gamma\tilde{W}^T W + 2\Gamma\tilde{W}^T\tilde{W}$$

$$-\Gamma^2\|I-\alpha\cdot\sigma\sigma^T\|W^T W + 2\Gamma\|I-\alpha\cdot\sigma\sigma^T\|W^T\tilde{W}$$

$$-\Gamma^2\|I-\alpha\cdot\sigma\sigma^T\|\tilde{W}^T\tilde{W}\}$$

$$<-\frac{\|I-\alpha\cdot\sigma\sigma^T\|}{\alpha}\{\Gamma(2-\Gamma)\|I-\alpha\cdot\sigma\sigma^T\|\|\tilde{W}\|^2 - \Gamma^2\|I-\alpha\cdot\sigma\sigma^T\|W_M^2\}$$

$$<-\frac{\|I-\alpha\cdot\sigma\sigma^T\|^2}{\alpha}\{\Gamma(2-\Gamma)\|\tilde{W}\|^2 - \Gamma^2 W_M^2\}.$$

Bounding,

$$\Delta L < -r^T[2I - (3+\alpha\sigma^T\sigma)K_v^T K_v - (\alpha\sigma^T\sigma)^2 + 2\alpha\sigma^T\sigma K_v]r$$

$$+2(1+\alpha\sigma^T\sigma)r^T K_v^T(D+\varepsilon) + 4r^T K_v^T D + 2D^T D$$

$$+(1+\alpha\sigma^T\sigma)\left(1+\frac{1+\alpha\sigma^T\sigma}{\beta}\right)(D+\varepsilon)^T(D+\varepsilon) - 2\alpha\sigma^T\sigma r^T D$$

$$+\rho^{-1}(\alpha\sigma^T\sigma + \Gamma\|I-\alpha\sigma\sigma^T\|)^2[K_v r + (1+\beta^{-1}(1+\alpha\sigma^T\sigma))(D+\varepsilon)$$

$$-\beta^{-1}\Gamma\|I-\alpha\sigma\sigma^T\|W^T\sigma]^T$$

$$\cdot[K_v r + (1+\beta^{-1}(1+\alpha\sigma^T\sigma))(D+\varepsilon) - \beta^{-1}\Gamma\|I-\alpha\sigma\sigma^T\|W^T\sigma]$$

$$+\beta^{-1}\Gamma^2\|I-\alpha\sigma\sigma^T\|^2\sigma^T WW^T\sigma$$

$$-2\beta^{-1}(1+\alpha\sigma^T\sigma)\Gamma\|I-\alpha\sigma\sigma^T\|(D+\varepsilon)^T W^T\sigma$$

$$-2\Gamma\|I-\alpha\cdot\sigma\sigma^T\|r^T K_v^T W^T\sigma - 2\Gamma\|I-\alpha\cdot\sigma\sigma^T\|(D+\varepsilon)^T W^T\sigma$$

$$-\frac{\|I-\alpha\cdot\sigma\sigma^T\|^2}{\alpha}\Gamma(2-\Gamma)\|\tilde{W}\|^2 + \frac{\|I-\alpha\cdot\sigma\sigma^T\|^2}{\alpha}\Gamma^2 W_M^2,$$

$$\Delta L < -r^T[2I - (3+\alpha\sigma^T\sigma)K_v^T K_v - (\alpha\sigma^T\sigma)^2 + 2\alpha\sigma^T\sigma K_v]r$$

$$+2(1+\alpha\sigma^T\sigma)r^T K_v^T(D+\varepsilon) + 4r^T K_v^T D + 2D^T D$$

$$+(1+\alpha\sigma^T\sigma)[1+\beta^{-1}(1+\alpha\sigma^T\sigma)](D+\varepsilon)^T(D+\varepsilon) - 2\alpha\sigma^T\sigma r^T D$$

$$+ \rho^{-1}(\alpha\sigma^T\sigma + \Gamma\|I - \alpha\sigma\sigma^T\|)^2$$
$$\cdot [r^T K_v^T K_v r + (1 + \beta^{-1}(1 + \alpha\sigma^T\sigma))^2 (D + \varepsilon)^T (D + \varepsilon)$$
$$+ \beta^{-2}\Gamma^2\|I - \alpha\sigma\sigma^T\|^2 \sigma^T W W^T\sigma + 2(1 + \beta^{-1}(1 + \alpha\sigma^T\sigma))(D + \varepsilon)^T K_v r$$
$$- 2\beta^{-1}\Gamma\|I - \alpha\sigma\sigma^T\|\sigma^T W K_v r$$
$$- 2(1 + \beta^{-1}(1 + \alpha\sigma^T\sigma))\beta^{-1}\Gamma\|I - \alpha\sigma\sigma^T\|(D + \varepsilon)^T W^T\sigma]$$
$$+ \beta^{-1}\Gamma^2\|I - \alpha\sigma\sigma^T\|^2 \sigma^T W W^T\sigma$$
$$- 2\beta^{-1}(1 + \alpha\sigma^T\sigma)\Gamma\|I - \alpha\sigma\sigma^T\|(D + \varepsilon)^T W^T\sigma$$
$$- 2\Gamma\|I - \alpha\cdot\sigma\cdot\sigma^T\|r^T K_v^T W^T\sigma - 2\Gamma\|I - \alpha\cdot\sigma\cdot\sigma^T\|(D + \varepsilon)^T W^T\sigma$$
$$- \frac{\|I - \alpha\cdot\sigma\cdot\sigma^T\|^2}{\alpha}\Gamma(2 - \Gamma)\|\tilde{W}\|^2 + \frac{\|I - \alpha\cdot\sigma\cdot\sigma^T\|^2}{\alpha}\Gamma^2 W_M^2.$$

Define $\eta = (1 + \alpha\sigma^T\sigma)I + \rho^{-1}(\alpha\sigma^T\sigma + \Gamma\|I - \alpha\sigma\sigma^T\|)^2 > 0$ (condition (4.2.34)) and $\gamma = \eta + \beta^{-1}\rho^{-1}(1 + \alpha\sigma^T\sigma)(\alpha\sigma^T\sigma + \Gamma\|I - \alpha\sigma\sigma^T\|)^2 > 0$. Substituting,

$$\Delta J < -r^T[2I - (\alpha\sigma^T\sigma)^2 + 2\alpha\sigma^T\sigma K_v - (\eta + 2)K_v^T K_v]r$$
$$+ 2\gamma \cdot r^T K_v^T (D + \varepsilon) + \gamma[1 + \beta^{-1}(1 + \alpha\sigma^T\sigma)](D + \varepsilon)^T (D + \varepsilon)$$
$$+ 4r^T K_v^T D + 2D^T D - 2\alpha\sigma^T\sigma r^T D$$
$$- 2\Gamma\|I - \alpha\cdot\sigma\cdot\sigma^T\|[\beta^{-1}\rho^{-1}(\alpha\sigma^T\sigma + \Gamma\|I - \alpha\sigma\sigma^T\|)^2 + 1]\sigma^T W K_v r$$
$$- 2\Gamma\|I - \alpha\cdot\sigma\sigma^T\|[1 + \beta^{-1}(1 + \alpha\sigma^T\sigma)]$$
$$\cdot [1 + \beta^{-1}\rho^{-1}(\alpha\sigma^T\sigma + \Gamma\|I - \alpha\sigma\sigma^T\|)^2](D + \varepsilon)^T W^T\sigma$$
$$- \frac{\|I - \alpha\cdot\sigma\sigma^T\|^2}{\alpha}\Gamma(2 - \Gamma)\|\tilde{W}\|^2 + \frac{\|I - \alpha\cdot\sigma\sigma^T\|^2}{\alpha}\Gamma^2 W_M^2$$
$$+ \beta^{-2}\rho^{-1}(\alpha\sigma^T\sigma + \Gamma\|I - \alpha\sigma\sigma^T\|)^2\Gamma^2\|I - \alpha\sigma\sigma^T\|^2\sigma^T W W^T\sigma$$
$$+ \beta^{-1}\Gamma^2\|I - \alpha\sigma\sigma^T\|^2\sigma^T W W^T\sigma,$$
$$\Delta J < -[2 - (\alpha\sigma^T\sigma)^2 + 2\alpha\sigma^T\sigma K_{V\min} - (\eta + 2)K_{v\max}^2]\|r\|^2$$
$$+ 2\gamma \cdot \|r\|K_{v\max}(D_M + \varepsilon_M) + \gamma[1 + \beta^{-1}(1 + \alpha\sigma^T\sigma)](D_M + \varepsilon_M)^2$$
$$+ 4\|r\|K_{v\max}D_M + 2D_M^2 + 2\alpha\sigma_M^2\|r\|D_M$$
$$+ 2\Gamma\|I - \alpha\cdot\sigma\sigma^T\|[\beta^{-1}\rho^{-1}(\alpha\sigma^T\sigma + \Gamma\|I - \alpha\sigma\sigma^T\|)^2 + 1]\sigma_M W_M K_{v\max}\|r\|$$
$$+ 2\Gamma\|I - \alpha\cdot\sigma\sigma^T\|[1 + \beta^{-1}(1 + \alpha\sigma^T\sigma)]$$
$$\cdot [1 + \rho^{-1}\beta^{-1}(\alpha\sigma^T\sigma + \Gamma\|I - \alpha\sigma\sigma^T\|)^2](D_M + \varepsilon_M)W_M\sigma_M$$
$$- \frac{\|I - \alpha\cdot\sigma\sigma^T\|^2}{\alpha}\Gamma(2 - \Gamma)\|\tilde{W}\|^2 + \frac{\|I - \alpha\cdot\sigma\sigma^T\|^2}{\alpha}\Gamma^2 W_M^2$$
$$+ \rho^{-1}\beta^{-2}(\alpha\sigma^T\sigma + \Gamma\|I - \alpha\sigma\sigma^T\|)^2\Gamma^2\|I - \alpha\sigma\sigma^T\|^2\sigma_M^2 W_M^2$$
$$+ \beta^{-1}\Gamma^2\|I - \alpha\sigma\sigma^T\|^2\sigma_M^2 W_M^2.$$

Define $\rho_1 = 2 - (\alpha\sigma^T\sigma)^2 + 2\alpha\sigma^T\sigma K_{V\min} - (\eta + 2)K_{v\max}^2 > 0$ (see conditions (4.2.29) and (4.2.31)),

$$\rho_2 = \Gamma\|I - \alpha\cdot\sigma\sigma^T\|[\|\beta^{-1}\rho^{-1}\|(\alpha\sigma^T\sigma + \Gamma\|I - \alpha\sigma\sigma^T\|)^2 + 1]\sigma_M W_M K_{v\max}$$
$$+ \gamma K_{v\max}\varepsilon_M + [(\gamma + 2)K_{v\max} + \alpha\sigma_M^2]D_M,$$

$$\rho_3 = \gamma[1 + \|\beta^{-1}\|(1 + \alpha\sigma^T\sigma)](D_M + \varepsilon_M)^2 + 2D_M^2$$
$$+ 2\Gamma\|I - \alpha \cdot \sigma \cdot \sigma^T\|[1 + \|\beta^{-1}\|(1 + \alpha\sigma^T\sigma)]$$
$$\cdot [1 + \|\rho^{-1}\beta^{-1}\|(\alpha\sigma^T\sigma + \Gamma\|I - \alpha\sigma\sigma^T\|)^2] \cdot (D_M + \varepsilon_M)W_M\sigma_M$$
$$+ \frac{\|I - \alpha \cdot \sigma \cdot \sigma^T\|^2}{\alpha}\Gamma^2 W_M^2 + \|\beta^{-1}\|$$
$$\cdot \Gamma^2\|I - \alpha\sigma\sigma^T\|^2[1 + \|\rho^{-1}\beta^{-1}\|(\alpha\sigma^T\sigma + \Gamma\|I - \alpha\sigma\sigma^T\|)^2]\sigma_M^2 W_M^2,$$
$$\Delta J < -\rho_1\|r\|^2 + 2\rho_2\|r\| + \rho_3 - \frac{\|I - \alpha \cdot \sigma \cdot \sigma^T\|^2}{\alpha}\Gamma(2 - \Gamma)\|\tilde{W}\|^2.$$

Completing squares for $\|r\|$,

$$\Delta J < -\rho_1\left[\|r\| - \frac{\rho_2}{\rho_1}\right]^2 + \frac{\rho_2^2}{\rho_1} + \rho_3 - \frac{\|I - \alpha \cdot \sigma \cdot \sigma^T\|^2}{\alpha}\Gamma(2 - \Gamma)\|\tilde{W}\|^2,$$

which is negative as long as

$$\frac{\|I - \alpha \cdot \sigma \cdot \sigma^T\|^2}{\alpha}\Gamma(2 - \Gamma)\|\tilde{W}\|^2 > \frac{\rho_2^2}{\rho_1} + \rho_3$$

$$\Rightarrow \|\tilde{W}\| > \frac{1}{\|I - \alpha \cdot \sigma \cdot \sigma^T\|}\sqrt{\frac{\alpha(\rho_2^2 + \rho_1\rho_3)}{\rho_1\Gamma(2 - \Gamma)}}$$

or

$$\rho_1\left[\|r\| - \frac{\rho_2}{\rho_1}\right]^2 > \frac{\rho_2^2}{\rho_1} + \rho_3 \Rightarrow \|r\| > \frac{\rho_2 + \sqrt{\rho_2^2 + \rho_1\rho_3}}{\rho_1}.$$

From the above results, ΔL is negative outside a compact set. According to a standard Lyapunov theorem extension (Lewis, Abdallah, and Dawson (1993)), it can be concluded that the tracking error $r(k)$, the actuator error $\tilde{\tau}(k)$, and the NN weights estimates $\tilde{W}(k)$ are GUUB.   □

Chapter 6

Fuzzy Logic Control of Vehicle Active Suspension

In this chapter, we present a backstepping-based FL scheme for the active control of vehicle suspension systems using the two-degrees-of-freedom or quarter-vehicle model. The full dynamics of a novel hydraulic strut are considered. The servovalve dynamics are also included. An FL system is used to estimate the nonlinear hydraulic strut dynamics. The backstepping FL scheme is shown to give superior performance over passive suspension and active suspension control using conventional PID schemes. The FL system is adapted in such a way as to estimate online the unknown strut hydraulic dynamics and provide the backstepping loop with the desired servovalve positioning so that the scheme becomes adaptive, guaranteeing bounded tracking errors and parameter estimates. A rigorous proof of stability and performance is given and a simulation example verifies performance. Unlike standard adaptive backstepping techniques, no LIP assumptions are needed.

6.1 Introduction

The study of vehicle ride quality evaluates the passenger's response to road/terrain irregularities with the objective of improving comfort and road isolation while maintaining wheel–ground contact. Ride problems mainly arise from vehicle vibrations, which may be induced by a variety of sources including external factors, such as roadway roughness or aerodynamics forces, or internal forces produced by vehicle subsystems, such as the engine, powertrain, or suspension mechanisms. Suspension system designs are mostly based on ride analysis. Vehicle suspensions using various types of springs, dampers, and linkages with tailored flexibility in various directions have been developed over the last century, since the beginning of the automobile age. The simplest and most common types of suspensions are passive in the sense that no external sources of energy are required.

With the development of modern control theory and the concomitant development of inexpensive and reliable electronic devices, it has become clear that increased performance is theoretically possible for suspensions that can sense aspects of vehicle motion and produce forces or motions through actuators in ways impossible for conventional passive suspensions. In the automobile industry, active and semiactive suspensions have recently attracted increased attention because they can improve vehicle ride comfort and road-handling performance (Karnopp (1983), Eiler and Hoogterp (1994), Karnopp and Heess (1991), Yoshimura and Hayashi (1996)). The main advantage of employing an active suspension is the associated adaptation potential. The suspension characteristic can be adjusted, while driving, to match the profile of the road being traversed. An active suspension system should be able to provide different behavioral characteristics depending upon various road conditions and to do so without going beyond its travel limits.

Including the dynamics of the hydraulic system consisting of fluids, valves, pumps, and so forth complicates the active suspension control problem even further since it introduces nonlinearities to the system. It has been noted that the hydraulic dynamics and fast servovalve dynamics make controls design very difficult. In Engelman and Rizzoni (1993), it was shown that the actuator dynamics significantly change the vibrational characteristics of the vehicle system. It was shown in Alleyne, Liu, and Wright (1998) that using a force control loop to compensate for the hydraulic dynamics can destabilize the system. The full nonlinear control problem for active suspension has been investigated using several approaches, including optimal control based on a linearized model (Engelman and Rozzoni (1993)), adaptive nonlinear control (Alleyne and Hedrick (1992, 1995)), and adaptive H_∞ control using backstepping (Fukao, Yamawaki, and Adachi (1998)). These schemes use linear approximations for the hydraulic dynamics or they neglect the servovalve model dynamics in which a current or voltage is what ultimately controls the opening of the valve to allow flow of hydraulic fluid to or from the suspension system. A nonlinear backstepping approach including the servovalve dynamics has been proposed in Lin and Kanellakopoulos (1997) in which a nonlinear filter whose effective bandwidth depends on the magnitude of the suspension travel is used.

During the past several years, there has been great success in the design of nonlinear control systems due to the development of the backstepping technique (Kanellakopoulos, Kokotović, and Morse (1991)). Integrator backstepping is used to design controllers for systems with known nonlinearities not satisfying the so-called "matching conditions." Adaptive backstepping design can be used to handle systems with unknown nonlinearities. However, adaptive backstepping design for nonlinear control may dramatically increase the complexity of the controller. Moreover, several assumptions of LIP are needed, which may not be satisfied by actual systems.

The use of FL systems has accelerated in recent years in many areas, including feedback control. An FL approach to active suspension control was given in Yoshimura and Hayashi (1996). An FL approach for the active control of a hydropneumatic actuator was presented in Cai and Konic (1996). Particularly important in FL control are the *universal function approximation capabilities* of FL systems (see Chapter 1). Given these recent results, some rigorous design techniques for FL feedback control based on *adaptive control* approaches have now been given (Kosko (1992), Wang (1994)). FL systems offer significant advantages over adaptive control, including no requirement for LIP assumptions and no need to compute a regression matrix for each specific system.

In this chapter, we give a mathematically rigorous stability proof and design algorithm for an adaptive backstepping FL controller for a two-degrees-of-freedom vehicle suspension model (quarter-vehicle model) (Campos Portillo (2000)). The dynamics of a novel hydraulic strut are included. The servovalve dynamics are also included. An FL system in the feedback loop estimates the nonlinear hydraulic dynamics. The desired force generation for the backstepping loop is performed by an active filtered feedback design proposed in Ikenaga et al. (1999). This approach assumes that the actual strut force exerted is available either by direct measurements using force sensors or estimation from some other measurements such as car body acceleration, suspension travel, and so forth. The FL system is adapted in such a way as to estimate online the unknown hydraulic dynamics and provide the backstepping loop with the desired servovalve positioning. The major advantage over standard adaptive backstepping approaches is that no LIP assumptions are needed.

Actual implementation of active suspension control systems on vehicle testbeds is discussed by Ikenaga (2000).

6.2 System model and dynamics

In this section, we give the dynamics of the vehicle and active suspension system. To focus on the problem of controlling the suspension actuator dynamics, we use the so-called quarter-vehicle dynamics, which deal only with a single wheel.

Quarter-vehicle active suspension system

The model of the quarter-vehicle active suspension system used in this chapter is shown in Figure 6.2.1. The quarter-vehicle active suspension system represents a single wheel of a car in which the wheel and axle are connected to the quarter-portion of the car body through an active hydraulic strut. The tire is modeled as a simple spring without damping. The equations of motion for this system are given as

$$m_s \ddot{z}_s + m_s g - F + F_f = 0,$$
$$m_u \ddot{z}_u + m_u g + K_t(z_r - z_u) + F - F_f = 0,$$

(6.2.1)

where m_s and m_u are the masses of car body (sprung mass) and wheel (unsprung mass), z_s and z_u are the displacements of car body and wheel, g is gravitational acceleration, K_t is the spring constant of the tire, z_r is the terrain input disturbance, F is the force produced by the active hydraulic strut, and F_f is the friction.

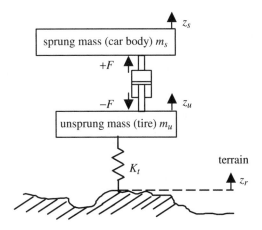

Figure 6.2.1. *Model of quarter-vehicle active suspension system.*

Active hydraulic strut

The active hydraulic strut used in this chapter is a patented compressible fluid suspension system developed by Davis Technologies International (DTI) in Dallas, Texas. The DTI strut offers significant advantages over other technologies in spring and damping performance, compact configuration, and overall system design/function flexibility. A model of the DTI active hydraulic strut is shown in Figure 6.2.2.

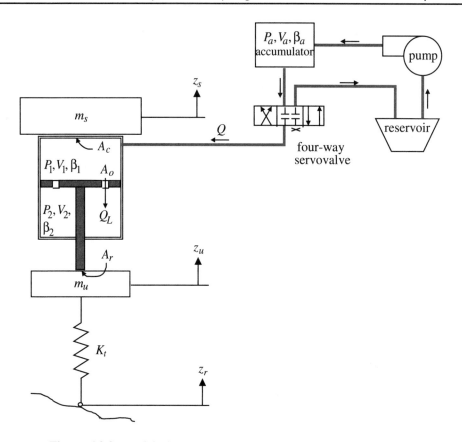

Figure 6.2.2. *Model of DTI's hydraulic strut with active force control.*

The dynamic equations for the DTI active hydraulic strut are given as

$$\dot{P}_1 = \beta_1 \left(\frac{Q - Q_L - A_c(\dot{z}_s - \dot{z}_u)}{V_{o1} + A_c(z_s - z_u)} \right),$$

$$\dot{P}_2 = \beta_2 \left(\frac{Q_L + (A_c - A_r)(\dot{z}_s - \dot{z}_u)}{V_{o2} - (A_c - A_r)(z_s - z_u)} \right), \tag{6.2.2}$$

where P_1 and P_2 are the pressures in chambers 1 and 2, A_c is the strut cylinder surface area, A_r is the piston rod surface area, A_0 is the piston hole surface area, and V_{01} and V_{02} are the initial volumes in chambers 1 and 2. Q, the compressible fluid flow between the accumulator and the strut, is modeled by the servovalve equation given by

$$Q = C_d w x_v \sqrt{\frac{2 |P_a \, \text{sat}(x_v) - P_1|}{\rho}}, \tag{6.2.3}$$

where C_d is the fluid discharge coefficient, ρ is the fluid density, w is the servovalve area gradient, P_a is the pressure in the accumulator (assumed to be constant), and x_v is the servovalve displacement from its "closed" position. The saturation function sat(.) is defined as

$$\text{sat}(x_v) = \begin{cases} 1, & x > 0, \\ 0, & x \le 0. \end{cases} \tag{6.2.4}$$

Q_L, the orifice flow through the piston hole area A_0, is computed as

$$Q_L = C_d A_0 \, \text{sgn}(P_1 - P_2) \sqrt{\frac{2|P_1 - P_2|}{\rho}}. \tag{6.2.5}$$

β_1 and β_2 are the bulk moduli for the fluid in chambers 1 and 2 and are given by

$$\beta_i(P_i) = 171617.03249\sqrt{1 + (7.718658 \times 10^{-5} P_i)^2}, \quad i = 1, 2. \tag{6.2.6}$$

This relationship was derived from experimental data.

The control input is the servovalve voltage $u(t)$ in

$$\tau \dot{x}_v = -x_v + u \tag{6.2.7}$$

with τ a time constant.

Full-state model of vehicle plus suspension system

Assigning the states as $x_1 = z_s$, $x_2 = \dot{z}_s$, $x_3 = z_s - z_u$, $x_4 = \dot{z}_u$, $x_5 = P_2$, $x_6 = P_1$, and $x_7 = x_v$ and combining all the equations, one obtains the state model for the vehicle plus strut dynamics as

$$
\begin{aligned}
\dot{x}_1 &= x_2, \\
\dot{x}_2 &= \frac{1}{m_s}(-m_s g + F - F_f), \\
\dot{x}_3 &= x_2 - x_4, \\
\dot{x}_4 &= \frac{1}{m_u}(-m_u g + k_t(z_r - x_1 + x_3) - F + F_f), \\
\dot{x}_5 &= \beta_2\left(\frac{Q_L + (A_c - A_r)(x_2 - x_4)}{V_{02} - (A_c - A_r)x_3}\right), \\
\dot{x}_6 &= \beta_1\left(\frac{Q - Q_L - A_c(x_2 - x_4)}{V_{01} + A_c x_3}\right), \\
\dot{x}_7 &= \frac{(-x_7 + u)}{\tau}, \\
\dot{x}_8 &= w_b(x_2 - x_4 - x_8), \\
\dot{x}_9 &= w_k(x_3 - x_9), \\
\dot{x}_{10} &= e_f = F_{\text{des}} - F, \\
\dot{x}_{11} &= \eta(e_f - x_{11}),
\end{aligned}
\tag{6.2.8}
$$

where

$$
\begin{aligned}
Q &= C_d w x_7 \sqrt{\frac{2|P_a \, \text{sat}(x_7) - x_6|}{\rho}}, \\
Q_L &= C_d A_0 \, \text{sgn}(x_6 - x_5)\sqrt{\frac{2|x_6 - x_5|}{\rho}}, \\
\beta_1 &= 171617.03249\sqrt{1 + (7.718658 \times 10^{-5} x_6)^2}, \\
\beta_2 &= 171617.03249\sqrt{1 + (7.718658 \times 10^{-5} x_5)^2}.
\end{aligned}
\tag{6.2.9}
$$

The force produced by the active hydraulic strut was given by Ikenaga et al. (1999):

$$F = A_c x_6 - (A_c - A_r) x_5.$$
(6.2.10)

6.3   Backstepping-based fuzzy logic controller

In this section, we use backstepping (Kanellakopoulos, Kokotović, and Morse (1991)) to design a vehicle active suspension controller (Campos Portillo (2000)). Backstepping design has two steps. First, one must design the ideal desired force control input to the vehicle assuming there are no actuator dynamics. This functions as an outer feedback control loop. Then in the second step, one takes into account the actuator dynamics to design an inner control loop that provides an input to the actuator, which causes it to exert an actual force that is close to the desired force.

6.3.1   Outer loop design for ideal force input

Here we design the outer control loop whose function is to produce a desired strut force. In the next section, FL backstepping design is used to close an inner control loop. Stability proofs are given.

A major limiting role in active suspension control is played by the wheel frequency w_0. This is given approximately as

$$w_0 = \sqrt{\frac{K_t}{m_u}}.$$
(6.3.1)

The standard control input location for active suspension control, as shown in Figure 6.2.1, is a force input applied between the sprung and unsprung masses. It is simply not feasible to apply control inputs between the sprung mass and the ground. Unfortunately, it is not possible to decrease body motions at the wheel frequency w_0 using control inputs applied only between the sprung and unsprung masses. This is due to the fact that using such control inputs, the system has a pole-zero cancellation at w_0 arising from an uncontrollable mode.

It is not difficult to improve performance below the wheel frequency w_0. This may be accomplished by rolling off the spring constant at low frequencies or using skyhook damping (Karnopp (1983), Ikenaga et al. (1999)). It is difficult to improve body motion at frequencies above the wheel frequency w_0 without using complex controllers such as backstepping (Alleyne and Hedrick (1995), Fukao, Yamawaki, and Adachi (1998), Lin and Kanellakopoulos (1997)) or mechanical modifications to the vehicle that include hanging extra dampers off the wheels. However, in Ikenaga et al. (1999) it was shown that rolling off the damping constants at high frequencies can easily mitigate motions above the wheel frequency.

It is possible to improve performance both above and below the wheel frequency w_0 by using active suspension control to introduce a damping constant B_s that is rolled off above w_0 and also a spring constant K_s that is rolled off below w_0 (Ikenaga et al. (1999)). This cannot be done using passive damping feedback but is possible with the DTI strut, which is a damper, a spring, and active suspension all in one device.

To roll off B_s at high frequency, one may introduce a low-pass filter (LPF) by defining

$$\bar{x}_{su} = \frac{w_b}{s + w_b} (\dot{z}_s - \dot{z}_u),$$
(6.3.2)

where w_b is the roll-off frequency for the damping constant B_s. This can be realized adding the state equation

$$\dot{\bar{x}}_{su} = -w_b \bar{x}_{su} + w_b (z_s - z_u). \tag{6.3.3}$$

To roll off K_s at low frequency, one may introduce a high-pass filter (HPF) by defining

$$x_{su} = \frac{s}{s + w_k}(z_s - z_u), \tag{6.3.4}$$

where w_k is the roll-off frequency for the spring constant K_s. This HPF is a washout circuit like those used in aircraft control. To realize the HPF as a state system, one may write

$$x_{su} = \frac{s}{s + w_k}(z_s - z_u) = \left(1 - \frac{w_k}{s + w_k}\right)(z_s - z_u) \tag{6.3.5}$$

and define an additional state so that

$$\dot{\tilde{x}} = -w_k \tilde{x} + w_k (z_s - z_u),$$
$$x_{su} = (z_s - z_u) - \tilde{x}. \tag{6.3.6}$$

The complete description of this system with vehicle mechanical dynamics, hydraulic dynamics, servovalve dynamics, and control system dynamics is given by the state equations (6.2.8), (6.3.3), and (6.3.6).

The overall control system design is shown in Figure 6.3.1. The outer feedback loop includes damping roll off, spring constant roll off, skyhook damping, and an additional ride error term e_{rh}. It produces the *ideal desired strut force* defined as

$$F_d = K_{rh} e_{rh} - K_s x_{su} - B_s \bar{x}_{su} - B_{sky} \dot{z}_s, \tag{6.3.7}$$

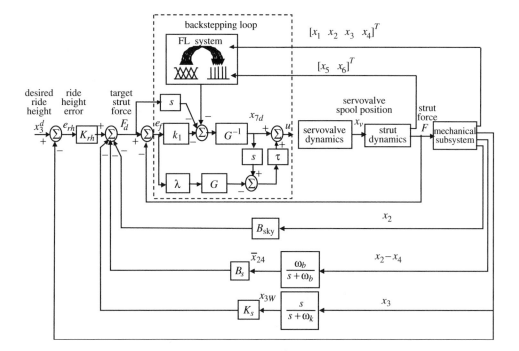

Figure 6.3.1. *Backstepping FL controller for the quarter-vehicle.*

where $e_{rh} = (z_s - z_u)_{\text{des}} - (z_s - z_u)$ and K_{rh} is the ride height error gain. Skyhook damping (Karnopp (1983)) is the term used to describe feedback of absolute velocity of the car body $\dot{z}_s$. The constants K_{rh} and B_{sky}, the nominal damping coefficient B_s, and the nominal spring constant K_s are design parameters selected by the control engineer based on computer simulation and vehicle tests.

This ideal desired control force has an error term involving e_{rh} that allows one to select the quarter-vehicle ride height. This is important in pitch/roll/heave control of the entire vehicle considering all four wheels (Campos Portillo (2000)).

Figure 6.3.2 shows the Bode magnitude plot of car body acceleration $\ddot{z}_s$ assuming the road disturbance $z_r(t)$ as the input. Only the vehicle mechanical subsystem (6.2.1) is considered. The solid plot is for passive suspension with constant damping equal to B_s and spring constant equal to K_s. The peak of vibration occurs at the wheel frequency w_0. The lower frequency peak is the vehicle body vibration frequency. The dotted Bode plot shows the effect of rolling off B_s above the wheel frequency and rolling off K_s below the wheel frequency. One sees that the response is significantly decreased both below and above the wheel frequency. Particularly note that the vibrations at the body frequency are greatly reduced. Note that, in keeping with the fundamental limitation of suspension control, the performance at the wheel frequency cannot be improved by applying forces only between the sprung and unsprung masses.

This demonstrates that the ideal force (6.3.7) improves the ride quality of the mechanical vehicle subsystem. The task in the next section is to add the strut dynamics and design a strut control outer loop that causes the strut to exert the ideal force (6.3.7) on the vehicle.

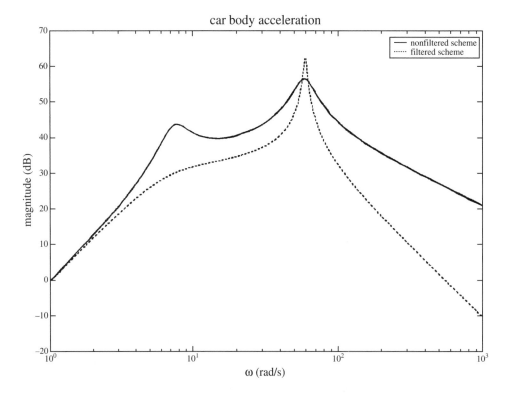

Figure 6.3.2. *Bode plot for mechanical subsystem with passive suspension (solid) and active filtered suspension control (dotted).*

6.3.2   Inner backstepping loop design

We have just specified an ideal desired force (6.3.7) that would cause the vehicle to have desirable properties. Now we complete the backstepping procedure by taking into account the actuator dynamics to design an inner control loop that provides an input to the actuator, which causes it to exert an actual force that is close to the desired force. An FL system simplifies the backstepping design by avoiding LIP assumptions. This also makes the design more relevant to actual physical systems, which may not satisfy LIP.

Define the force tracking error as

$$e_f = F_d - F = F_d - A_c x_6 + (A_c - A_r)x_5, \tag{6.3.8}$$

where F_d is (6.3.7) manufactured by the outer control loop. Differentiating,

$$\dot{e}_f = \dot{F}_d - \dot{F} = \dot{F}_d - A_c \dot{x}_6 + (A_c - A_r)\dot{x}_5. \tag{6.3.9}$$

Substituting from (6.2.8),

$$
\begin{aligned}
\dot{e}_f &= \dot{F}_d - A_c \beta_1 \left(\frac{Q - Q_L - A_c(x_2 - x_4)}{V_{01} + A_c x_3} \right) \\
&\quad + (A_c - A_r)\beta_2 \left(\frac{Q_L + (A_c - A_r)(x_2 - x_4)}{V_{02} - (A_c - A_r)x_3} \right) \\
&= \dot{F}_d + A_c \beta_1 \left(\frac{Q_L + A_c(x_2 - x_4)}{V_{01} + A_c x_3} \right) \\
&\quad + (A_c - A_r)\beta_2 \left(\frac{Q_L + (A_c - A_r)(x_2 - x_4)}{V_{02} - (A_c - A_r)x_3} \right) - \frac{Q A_c \beta_1}{V_{01} + A_c x_3}.
\end{aligned}
\tag{6.3.10}
$$

Using the expression for Q from (6.2.9), the above equation can be rewritten as

$$\dot{e}_f = \dot{F}_d + H(x_1, \dots, x_6) + G(x_1, \dots, x_6)x_7, \tag{6.3.11}$$

where

$$
\begin{aligned}
H(x_1, \dots, x_6) &= A_c \beta_1 \left(\frac{Q_L + A_c(x_2 - x_4)}{V_{01} + A_c x_3} \right) \\
&\quad + (A_c - A_r)\beta_2 \left(\frac{Q_L + (A_c - A_r)(x_2 - x_4)}{V_{02} - (A_c - A_r)x_3} \right),
\end{aligned}
\tag{6.3.12}
$$

$$G(x_1, \dots, x_6) = - \left(\frac{A_c \beta_1}{V_{01} + A_c x_3} \right) C_d w \sqrt{\frac{2|P_a \operatorname{sat}(x_7) - x_6|}{\rho}}.$$

$H(x_1, \dots, x_6)$ is a nonlinear function of the strut dynamics, which can be estimated by an FL system.

To compensate for the strut dynamics using the backstepping approach, we proceed in two phases as follows.

Phase 1. Select the desired servovalve displacement as

$$x_{7d} = -G^{-1}(\hat{H} + k_1 e_f + \dot{F}_d), \tag{6.3.13}$$

where $\hat{H}$ is the function $H(x_1, \dots, x_6)$ estimate and k_1 is a design parameter. $\hat{H}$ can be written as $\hat{H} = \hat{\theta}^T \xi(X)$, where $\hat{\theta} \in R^{q \times m}$ is a matrix of control representative values and $\xi(X) \in R^m$ is a vector of fuzzy basic functions (FBFs) with $X \in R^n$.

Then the force error dynamics can be written as

$$
\begin{aligned}
\dot{e}_f &= \dot{F}_d + H - GG^{-1}(\hat{H} + k_1 e_f + \dot{F}_d) + \dot{F}_d + H + Gx_7 - \dot{F}_d - H - Gx_{7d} \\
&= H - \hat{H} - k_1 e_f - Ge,
\end{aligned}
\tag{6.3.14}
$$

where $e = x_{7d} - x_7$ is the servovalve positioning error.

Phase 2. Define $e = x_{7d} - x_7$. Differentiate and use (6.2.8) to get

$$
\dot{e} = \dot{x}_{7d} - \dot{x}_7 = \dot{x}_{7d} - \frac{u - x_7}{\tau}.
\tag{6.3.15}
$$

Now select the servovalve control signal as

$$
u = x_{7d} + \tau\{\dot{x}_{7d} - \lambda G e_f\},
\tag{6.3.16}
$$

where λ is a design parameter and τ is the servovalve time constant.

Based on the FL approximation property, the actual nonlinear function can be represented as

$$
H = \theta^T \xi(X) + \varepsilon(X),
\tag{6.3.17}
$$

where $\varepsilon \le \varepsilon_N$ represents the FL approximation error which is a bounded and compact set by a known positive constant ε_N. Then the overall error dynamics will be

$$
\begin{aligned}
\dot{e}_f &= H - \hat{H} - k_1 e_f - Ge = (\theta^T - \hat{\theta}^T)\xi + \varepsilon - k_1 e_f - Ge \\
&= \tilde{\theta}^T \xi - k_1 e_f - Ge + \varepsilon, \\
\dot{e} &= \lambda G e_f - \frac{e}{\tau},
\end{aligned}
\tag{6.3.18}
$$

where $\tilde{\theta} = \theta - \hat{\theta}$ is the FL parameter error.

Some required mild assumptions are now stated. The two assumptions will be true in every practical situation and are standard in the existing literature.

Assumption 1 (Bounded ideal weights). The ideal values for the matrix of control representative values θ are bounded by known positive values so that

$$
\|\theta\|_F \le \theta_M.
\tag{6.3.19}
$$

Assumption 2 (Bounded desired trajectory). The desired trajectory is bounded in the sense, for instance, that

$$
\left\| \begin{matrix} x_d \\ \dot{x}_d \\ \ddot{x}_d \end{matrix} \right\| \le X_d.
\tag{6.3.20}
$$

The next theorem is our main result and shows how to tune the FL control representative values for the FL system to guarantee closed-loop stability.

Theorem 6.3.1 (Stability and tuning of FL backstepping suspension controller). *Consider the system given by (6.2.8). Let the control action $u(t)$ be provided by (6.3.16) with design parameter $\lambda > 0$. Let the desired servovalve position be provided by*

$$
x_{7d} = -G^{-1}(\hat{H} + k_1 e_f + \dot{F}_d)
\tag{6.3.21}
$$

with $\hat{H} = \hat{\theta}^T \xi(X)$ being the output of an FL system and $k_1 > 0$ being a design parameter. Let the tuning for the FL control representative values be given by

$$\dot{\hat{\theta}} = \Gamma\xi \cdot e_f - k_\theta \Gamma \|e_f\|\hat{\theta} \qquad (6.3.22)$$

with $\Gamma = \Gamma^T > 0$ and $k_\theta > 0$. Then the errors e_f and $\tilde{\theta}$ are UUB. Moreover, the force tracking error e_f can be arbitrarily small by increasing the fixed control gains k_1.

Proof. Let us consider the following Lyapunov function candidate:

$$L = \frac{1}{2}e_f^2 + \frac{1}{2\lambda}e^2 + \frac{1}{2}\mathrm{tr}[\tilde{\theta}^T \Gamma^{-1}\tilde{\theta}].$$

Differentiating and using the expression in (6.3.18),

$$\dot{L} = e_f\dot{e}_f + \frac{e\dot{e}}{\lambda} + \mathrm{tr}[\tilde{\theta}^T \Gamma^{-1}\dot{\tilde{\theta}}]$$

$$= e_f\tilde{\theta}^T\xi - k_1 e_f^2 - Ge_f e + e_f\varepsilon + eGe_f - \frac{e^2}{\lambda\tau} + \mathrm{tr}[(\theta - \hat{\theta})^T\Gamma^{-1}(\dot{\theta} - \dot{\hat{\theta}})]$$

$$= e_f\tilde{\theta}^T\xi - k_1 e_f^2 + e_f\varepsilon - \frac{e^2}{\lambda\tau} + \mathrm{tr}[(\hat{\theta} - \theta)^T\Gamma^{-1}\dot{\hat{\theta}}]$$

$$= e_f\tilde{\theta}^T\xi - k_1 e_f^2 + e_f\varepsilon - \frac{e^2}{\lambda\tau} + \mathrm{tr}[-\tilde{\theta}^T\Gamma^{-1}\dot{\hat{\theta}}].$$

Using the tuning law given by (6.3.22),

$$\dot{L} = e_f\tilde{\theta}^T\xi - k_1 e_f^2 + e_f\varepsilon - \frac{e^2}{\lambda\tau} + \mathrm{tr}[-\tilde{\theta}^T\Gamma^{-1}(\Gamma\xi \cdot e_f - k_\theta\Gamma\|e_f\|\hat{\theta})]$$

$$= e_f\tilde{\theta}^T\xi - e_f\tilde{\theta}^T\xi - k_1 e_f^2 + e_f\varepsilon - \frac{e^2}{\lambda\tau} + \mathrm{tr}[\tilde{\theta}^T k_\theta\|e_f\|\hat{\theta}]$$

$$= -k_1 e_f^2 + e_f\varepsilon - \frac{e^2}{\lambda\tau} + k_\theta\|e_f\|\mathrm{tr}[\tilde{\theta}^T(\theta - \tilde{\theta})]$$

$$\leq -k_1\|e_f\|^2 + e_f\varepsilon - \frac{e^2}{\lambda\tau} + k_\theta\|e_f\| \cdot \|\tilde{\theta}\| \cdot (\theta_M - \|\tilde{\theta}\|)$$

$$= -\|e_f\| \cdot [k_1\|e_f\| - \varepsilon - k_\theta\|\tilde{\theta}\|(\theta_M - \|\tilde{\theta}\|)] - \frac{e^2}{\lambda\tau}.$$

Completing squares for the term inside the brackets yields

$$k_1\|e_f\| - \varepsilon + k_\theta\|\tilde{\theta}\|^2 - k_\theta\|\tilde{\theta}\|\theta_M = k_\theta\left(\|\tilde{\theta}\| - \frac{\theta_M}{2}\right)^2 + k_1\|e_f\| - \varepsilon - \frac{k_\theta\theta_M^2}{4},$$

which is positive (i.e., $\dot{L} < 0$) as long as

$$\|e_f\| \geq \frac{\varepsilon + \frac{k_\theta\theta_M^2}{4}}{k_1} \qquad (6.3.23)$$

or

$$\|\tilde{\theta}\| > \frac{\theta_M}{2} + \sqrt{\frac{\varepsilon}{k_\theta} + \frac{\theta_M^2}{4}}. \qquad (6.3.24)$$

Therefore, $\dot{L}$ is negative outside a compact set. According to a standard Lyapunov theory extension, this demonstrates the UUB of the errors e_f and $\tilde{\theta}$.   □

This theorem shows the structure of the FL controller as derived using backstepping. It also shows how to tune the FL weights so that stability is guaranteed. Equation (6.3.23) provides a practical bound on the force generation error in the sense that e_f will never stray far above this bound. Moreover (6.3.24) shows that the FL parameter estimation error is bounded, so that the control signal is bounded.

6.4   Simulation results

The values for the system parameters given in (6.2.8) were selected as (Ikenaga et al. (1999))

$$
\begin{aligned}
m_s &= 290 \text{Kg}, & A_c &= 5.35225 \times 10^{-3} \text{m}^2, \\
m_u &= 59 \text{Kg}, & A_0 &= 3.35483 \times 10^{-4} \text{m}^2, \\
K_t &= 190000 \text{N/m}, & A_r &= 3.87741 \times 10^{-4} \text{m}^2, \\
C_d &= 0.7, & \rho &= 970 \text{Kg/m}^3, \\
w &= 1.43633 \times 10^{-2} \text{m}^2, & P_a &= 34473789.5 \text{Pa}.
\end{aligned}
$$

The wheel frequency for this system is approximately given by

$$
w_0 = \sqrt{\frac{K_t}{m_u}} = 56.75 \text{rad/s} \quad \text{or} \quad 9.03 \text{Hz}.
$$

It is important to note the units of the design parameters, $k_1 \to \text{s}^{-1}$ (Hz) and $\lambda \to \text{m}^2/\text{N}^2$. The values for the feedback damping, spring, and skyhook constants were selected as $K_{rh} = 16812 \text{N/m}$, $K_s = 16812 \text{N/m}$, $B_s = 1000 \text{N/m/sec}$, and $B_{\text{sky}} = 2000 \text{N/m/sec}$. The road disturbance was selected as $z_r = 0.04 \sin(w_r t) \text{m}$.

In this example, the system was simulated for three different values of frequency for the terrain disturbance z_r. We selected $w_r = 8 \text{rad/sec}$, $w_r = 58 \text{rad/sec}$, and $w_r = 150 \text{rad/sec}$. The value of $w_r = 58 \text{rad/sec}$ is very close to the wheel frequency for this example.

Figure 6.4.1 shows the results of the payload velocity with passive suspension only. In this case there are no physical restrictions on the strut travel. In practical situations, there will be limitations on the strut travel. Imposing these limitations on the passive suspension case, we get the results shown in Figure 6.4.2. It can be seen that the results at the two higher frequencies are much worse than in the case with no restrictions. Thus strut travel limitations significantly degrade the performance of passive suspension systems.

Next, we simulated an active suspension controller that uses only a naïve design using PID control. The results are shown in Figure 6.4.3. The strut travel limits are included here. Even though the amplitude is reduced at low frequencies by using the PID controller, a second high-frequency vibratory mode appears on the low-frequency plot. It turns out that this mode comes from the servovalve dynamics. It can be seen that this PID control system becomes unstable at the wheel frequency w_0. At higher frequencies, the amplitude is reduced over that accomplished with passive suspension control, though the high-frequency spike component might prove objectionable to the passenger.

It is clear that if there are servovalve dynamics, one must take some care in designing active suspension systems. In fact, a naïve design can destabilize the system for road disturbances occurring at the wheel frequency. Therefore, we next simulate the performance using

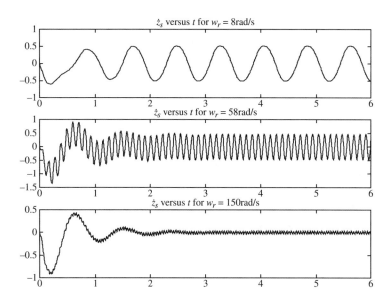

Figure 6.4.1. *Simulation results with passive suspension (no physical limits).*

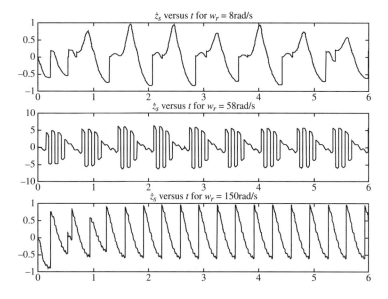

Figure 6.4.2. *Simulation results with passive suspension (with physical limits).*

the proposed backstepping-based FL controller. Figure 6.4.4 shows the results with the strut travel limits again included. An overall improvement can be seen with respect to both the passive suspension design and the active design using the PID controller. At low frequencies, the car-body motion amplitude is reduced and the resulting velocity is smoother. This is due to the fact that the FL backstepping controller suppresses the servovalve vibratory mode. The FL backstepping controller also improves the performance in the presence of strut travel limits. At the wheel frequency, the response is stable but there is not much improvement.

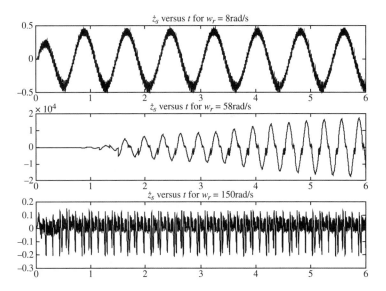

Figure 6.4.3. *Simulation results with PID controller.*

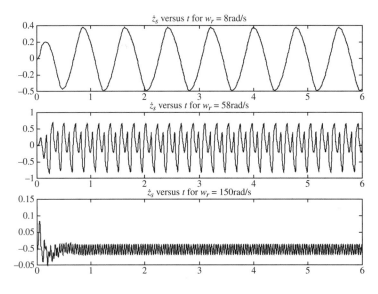

Figure 6.4.4. *Simulation results with the backstepping FL controller.*

This is expected due to the fundamental limitation of the invariant behavior at the wheel frequency for suspensions that exert forces between the sprung and unsprung masses. The performance at higher frequencies is much better than using passive suspension.

Chapter 7

Neurocontrol Using the Adaptive Critic Architecture

This chapter is concerned with the application of adaptive critic techniques to feedback control of nonlinear systems using NNs. No initial model for the nonlinear system is necessary. This work shows how to cope with nonlinearities through adaptive critics with no preliminary offline learning required. The proposed scheme consists of a feedforward action-generating NN that compensates for the unknown system nonlinearities. The learning of this NN is performed online based on a signal from a second higher level NN acting as a critic. The critic NN tunes itself online using a certain performance measure of the system. Both NN tuning algorithms are based on backpropagation, which must be modified to guarantee closed-loop performance. The two algorithms are derived using nonlinear stability analysis, so that both system tracking stability and error convergence can be guaranteed in the closed-loop system. This chapter brings previous work done in NN control closer towards the structure of the human cerebello–rubrospinal system and improves the learning capability of the NN controller.

7.1 Introduction

NNs have been shown to be very effective for the control of nonlinear dynamical systems. Rigorous design techniques and stability proofs for basic NN feedback controllers have by now been given (Chen and Khalil (1992), Lewis, Yesildirek, and Liu (1996), Narendra and Parthasarathy (1990), Polycarpou (1996)). NNs in the basic feedback control loop use the learning ability, nonlinear parallel processing ability, and function approximation capability of NNs (see Chapter 1). It is desired to extend the uses of NNs in systems design upwards from the basic feedback control loops. Major topics of study towards this goal would include supervised learning (Werbos (1992)), inverse control, neural adaptive control, backpropagation of utility (Werbos (1974)), and adaptive critic architectures (Gullapalli, Franklin, and Benbrahuim (1994), Kim and Lewis (1997, 1998), Werbos (1992), Yang and Asada (1993)).

NN architectures and learning are discussed in Bose and Liang (1996) and Haykin (1994). Numerous higher level architectures for NN control systems have been proposed in the literature. Excellent references include Miller, Sutton, and Werbos (1991) and White and Sofge (1992).

NN learning methods have been divided into three main paradigms: unsupervised learning, supervised learning, and reinforcement learning. The unsupervised learning method does not require an external teacher or supervisor to guide the learning process. Instead, the teacher can be considered to be built into the learning method. Unlike the unsupervised learning method, both supervised and reinforcement learning require an external teacher to provide training signals that guide the learning process. The difference between these two paradigms arises from the kind of information about the local characteristics of the performance surface that is available to the learning system (provided by the teacher). It is useful to discuss how supervised learning and reinforcement learning methods differ. In a supervised learning task, information is available about the gradient of the performance surface at the point corresponding to the system's current behavior. This gradient information often takes the form of an error vector giving the difference between the system's action and some desired, or target action. This is the case, for example, in systems learning from examples, such as pattern classification and function approximation systems. The key point is that in supervised schemes, the learning is provided with information about whether or not a local improvement is possible and, if possible, which behaviors should be changed to achieve improvement.

In reinforcement learning, on the other hand, the role of the teacher is more evaluative than instructional. Instead of directly receiving detailed gradient information during learning, the learning system receives only information about the current value of a system performance measure. In contrast to information about a gradient, information about the performance measure value does not itself indicate how the learning system should change its behavior to improve performance; there is no directed information. A system employing reinforcement learning has to concern itself with estimating gradients based on information about the values that the performance measure takes over time. The key point is that information available in reinforcement learning evaluates behavior but does not in itself indicate if improvement is possible or how to make direct changes in behavior.

Because detailed knowledge of the controlled system and its behavior is not needed, reinforcement learning is potentially one of the most useful NN approaches to feedback control systems. Reinforcement learning is based on the common sense notion that if an action is followed by a satisfactory state of affairs, or by an improvement in the state of affairs, then the tendency to produce that action should be strengthened (i.e., reinforced). Extending this idea to allow action selections to depend on state information introduces aspects of feedback control, pattern recognition, and associative learning.

The idea of "adaptive critic" is an extension of this general idea of reinforcement. The signal provided by the critic in adaptive critic schemes conveys much less information than the desired output required in supervised learning. Nevertheless, their ability to generate correct control actions makes adaptive critics important candidates, where the lack of detailed structure in the task definition makes it difficult to define a priori the desired outputs for each input, as required by supervised learning control and direct inverse control.

The adaptive critic NN architecture uses a high-level supervisory NN that critiques system performance and tunes a second NN in the feedback control loop, which is known as the action-generating loop. This two-tier structure is based on human biological structures in the cerebello–rubrospinal system.

Adaptive critics have been used in an ad hoc fashion in NN control. No proofs acceptable to the control community have been offered for the performance of this important structure for feedback control. In standard control theory applications with NNs, the NN was used in the action-generating loop (i.e., basic feedback control loop) to control a system. Tuning laws for that case were given that guarantee stability and performance. However, a high-level critic was not used.

In this chapter, we propose a new type of adaptive critic useful in feedback control that is based on a critic signal provided by an NN. For the first time we are aware of, we offer a rigorous analysis of the adaptive critic structure, providing structural modifications and tuning algorithms required to guarantee closed-loop stability. This provides a connection between the work in NN feedback control and the structure of the human cerebello–rubrospinal system. The adaptive critic architecture also improves the learning capability of the NN controller. In fact, the adaptive critic architecture is used to overcome the effects of the system nonlinearities and external disturbances on the performance in a closed loop. The NN learning is performed online as the system is controlled, with no offline learning phase required. Closed-loop performance is guaranteed through the learning algorithms proposed.

7.2 Dynamics of a nonlinear system

Industrial processes are complex systems with nonlinear dynamics. Many industrial processes have dynamics of the general nonlinear form

$$\dot{x}_1 = x_2,$$

$$\vdots$$

$$\dot{x}_{n-1} = x_n, \tag{7.2.1}$$

$$\dot{x}_n = f(x) + u(t) + d(t),$$

$$y = x_1,$$

with state $x = [x_1 \quad x_2 \quad \cdots \quad x_n]^T$, $u(t)$ the control input to the plant, and $y(t)$ the output of interest. Signal $d(t)$ denotes the unknown disturbance, which we assume has a known upper bound b_d. The system nonlinearities are given by the smooth function $f(x) : \Re^n \rightarrow \Re^m$. It is assumed that the nonlinearity $f(x)$ in the system and the external disturbances $d(t)$ are unknown to the controller. System (7.2.1) is known as an mnth-order MIMO system in Brunovsky form.

Given a desired trajectory and its derivative values

$$x_d(t) = [x_d \quad \dot{x}_d \quad \ldots \quad x_d^{(n)}], \tag{7.2.2}$$

define the tracking error as

$$e(t) = x(t) - x_d(t), \tag{7.2.3}$$

which captures the performance of the closed-loop system state $x(t)$ in tracking the desired trajectory $x_d(t)$.

It is typical in robotics to define a so-called filtered tracking error $r(t) \in R^m$ as

$$r(t) = e^{(n-1)}(t) + \lambda_{n-2} e^{(n-2)}(t) + \cdots + \lambda_1 e^{(1)}(t) + e(t), \tag{7.2.4}$$

or in matrix form,

$$r(t) = [\Lambda^T \quad 1] \cdot \bar{e}(t), \tag{7.2.5}$$

where $e^{(n-1)}(t), \ldots, e^{(1)}(t)$ are the derivative values of the error $e(t)$,

$$\bar{e}(t) = [e^{(n-1)}(t) \quad e^{(n-2)}(t) \quad \cdots \quad e^{(1)}(t) \quad e(t)]^T,$$

and $\Lambda = [\lambda_{n-2} \quad \lambda_{n-3} \quad \cdots \quad \lambda_1]^T$ is an appropriate chosen vector of constant values so that $|s^{n-1} + \lambda_{n-2} s^{n-2} + \cdots + \lambda_1|$ is stable (i.e., Hurwitz). This means that $e(t) \rightarrow 0$ exponentially as $r(t) \rightarrow 0$. The performance measure $r(t)$ can be viewed as the real-valued instantaneous utility function of the plant performance. When $r(t)$ is small, system performance is good.

7.3   Adaptive critic feedback controller

For nonlinear systems, even if a good model of the nonlinear system is known, it is often difficult to formulate control laws. If unknown nonlinearities like friction, backlash, dead-zone, and external disturbances exist, the problem is even more difficult. We now show that the adaptive critic NN architecture gives a control system that effectively guarantees good tracking performance and disturbance rejection by actually learning the system nonlinearities and disturbances as time passes.

7.3.1   Adaptive critic neurocontrol architecture

Figure 7.3.1 shows the architecture of the proposed adaptive critic neurocontroller, showing the overall adaptive scheme whose details are subsequently derived. The adaptive critic neurocontroller designed in this chapter is a combination of an action-generating NN (labeled NN2 in the figure) that produces the control input for the system, a higher level NN (labeled NN1 in the figure) that provides an adaptive-learning signal, and a fixed gain controller $K_v r(t)$ in the performance measurement loop which uses an error based on the given reference trajectory. The controller is designed with the objective of learning how to generate the best control action in the absence of complete information about the plant and external disturbances, or of any training sets.

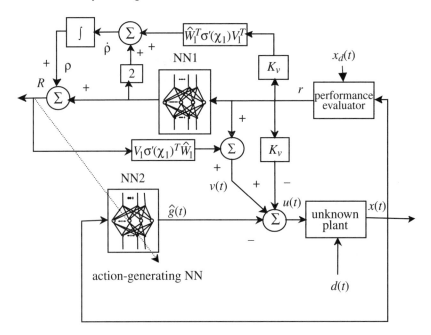

Figure 7.3.1. *Adaptive critic neurocontrol architecture.*

The behavior of the adaptive critic neurocontroller is that the performance measurement loop measures the system performance for the current system states assuming no disturbance or nonlinearity is present, and at the same time provides information to a second NN, which criticizes it and then supplies the learning signal for the action-generating NN. Then the action-generating NN generates the counter-signal necessary to overcome the nonlinearities and external disturbances, which the performance loop cannot deal with.

Some required mild assumptions are now stated. The two assumptions will be true in every practical situation and are standard in the existing literature.

Assumption 1. The ideal NN weights W are bounded by known positive values so that the Frobenius norm (Chapter 2) satisfies

$$\|W\|_F \leq W_M. \tag{7.3.1}$$

Assumption 2. The desired trajectory is bounded in the sense, for instance, that

$$\left\| \begin{matrix} q_d \\ \dot{q}_d \\ \ddot{q}_d \end{matrix} \right\| \leq Q_d. \tag{7.3.2}$$

A choice for the critic signal R is

$$R = \hat{W}_1^T \cdot \sigma(\chi_1) + \rho, \tag{7.3.3}$$

where ρ is an adaptive term that will be detailed later and the first term is the output of NN1, the critic NN. The current weights are denoted as $\hat{W}_1$ and the input vector to the NN1 hidden neuron layer as χ_1.

Using equations (7.2.1)–(7.2.3), the dynamics of the performance measure signal (7.2.4) can be written as

$$\dot{r} = g(x, x_d) + u(t) + d(t), \tag{7.3.4}$$

where $g(x, x_d)$ is a complex nonlinear function of the state and desired trajectory vectors x and x_d, respectively. Note that this function includes the original system unknown nonlinear function $f(x)$.

According to the approximation properties of NNs, the continuous nonlinear function $g(x, x_d)$ in (7.3.4) can be represented as

$$g(x, x_d) = W_2^T \sigma(\chi_2) + \varepsilon(\chi_2), \tag{7.3.5}$$

where χ_2 is the input vector to the NN2 hidden neurons and the NN reconstruction error $\varepsilon(\chi_2)$ is bounded by a known constant ε_N. The ideal weights W_2 that approximate $g(\cdot)$ are unknown.

Let the NN functional estimate for the continuous nonlinear function $g(x, x_d)$ be given by a second action-generating NN as

$$\hat{g}(x, x_d) = \hat{W}_2^T \sigma(\chi_2), \tag{7.3.6}$$

where $\hat{W}_2$ are the current weights estimating W_2.

From the adaptive critic neurocontroller architecture, the control input $u(t)$ is given by

$$u(t) = -K_v r - \hat{g}(x, x_d) + v(t), \tag{7.3.7}$$

where $\hat{g}(x, x_d)$ is provided by the action-generating NN, the control gain matrix is $K_v = K_v^T > 0$, and $v(t)$ is a robustifying vector that will be used to offset the NN functional reconstruction error $\varepsilon(\chi_2)$ and disturbances $d(t)$. Using (7.3.7), we can rewrite the closed-loop performance measure dynamics (7.3.4) as

$$\dot{r} = -K_v r + \tilde{g}(x, x_d) + d(t) + v(t), \tag{7.3.8}$$

where the functional estimation error is defined as $\tilde{g}(x, x_d) = g(x, x_d) - \hat{g}(x, x_d)$.

Using (7.3.5) and (7.3.6) in (7.3.8), we have the following dynamics for the performance measure:

$$\dot{r} = -K_v r + \tilde{W}_2^T \sigma(\chi_2) + \varepsilon(\chi_2) + d(t) + v(t) \qquad (7.3.9)$$

with the weight estimation error $\tilde{W}_2 = W_2 - \hat{W}_2$.

The next theorem is our main result and shows how to adjust the weights of both NNs to guarantee closed-loop stability.

Theorem 7.3.1 (Adaptive critic neurocontroller). *Let the control action $u(t)$ be provided by (8.5.12) and the robustifying term be given by*

$$v(t) = -k_z \cdot \frac{V_1 \sigma'(\chi_1) \hat{W}_1 R + r}{\|V_1 \sigma'(\chi_1) \hat{W}_1 R + r\|} \qquad (7.3.10)$$

with $k_z \geq b_d$. Let the critic signal be provided by

$$R = \hat{W}_1^T \sigma(\chi_1) + \rho \qquad (7.3.11)$$

with $\hat{W}_1^T \sigma(\chi_1)$ being the output of a critic NN and ρ being an auxiliary adaptive term specified later. Let the tuning for the critic and action-generating NNs be

$$\dot{\hat{W}}_1 = -\sigma(\chi_1) R^T - \hat{W}_1, \qquad (7.3.12)$$

$$\dot{\hat{W}}_2 = \Gamma \sigma(\chi_2) \cdot (r + V_1 \sigma'(\chi_1)^T \hat{W}_1 R)^T - \Gamma \hat{W}_2 \qquad (7.3.13)$$

with $\Gamma = \Gamma^T > 0$. Finally, let the auxiliary adaptive term ρ be tuned by

$$\dot{\rho} = \hat{W}_1^T [2\sigma(\chi_1) + \sigma'(\chi_1) V_1^T K_v r]. \qquad (7.3.14)$$

Then the errors r, $\tilde{W}_1$, and $\tilde{W}_2$ are UUB. Moreover, the performance measure $r(t)$ can be arbitrarily small by increasing the fixed control gains K_v.

Proof. See the appendix of this chapter (section 7.A).

Only one parameter Γ, the training rate for the action-generating NN, must be selected. This makes the tuning process less complicated than other NN schemes. This is a major advantage, as one may gather by examining the discussion on parameter selection in the simulation section of previous chapters.

The adaptive critic neurocontroller does not require the linearity in the unknown system parameters that is required in standard adaptive control since the NN functional approximation property holds for all continuous functions over a compact set.

Due to the last term in the weight adaptation rules (7.3.12) and (7.3.13), the persistent excitation conditions usually required in standard adaptive control are not needed. It is worth mentioning that advanced adaptive control schemes (Ioannou and Datta (1991), Narendra and Annaswamy (1987)) use a deadzone, e-mod, or σ-mod terms in the tuning law to avoid the need of the persistent excitation condition.

Both NNs are tuned online as they control the system. No preliminary offline training is needed. It is straightforward to initialize the weights and have guaranteed stability, unlike other NN control techniques proposed in the literature. In fact, if one selects $\hat{W}_1 = 0$,

$\hat{W}_2 = 0$, then the performance evaluation loop functions as a PD feedback loop that keeps the plant stable until the NN begins to learn. As the NNs learn, the performance improves.

It is very interesting to note the relation between the critic NN, which has weights W_1, and the action-generating NN, which has weights W_2. Though the two NNs are used as distinct and separate networks, in the tuning algorithm (7.3.12), (7.3.13), they are coupled together. In fact, the first terms of (7.3.12), (7.3.13) are continuous-time versions of backpropagation (note the Jacobian appearing in (7.3.13)). Therefore, it appears that the critic NN is effectively performing as the second layer of a *single NN with two layers* of adjustable weights, which also contains the action-generating NN as layer 1. The philosophical ramifications of this are still under study.

7.4   Simulation results

In this section, the adaptive critic neurocontroller is simulated on a digital computer. It is found to be very efficient at canceling the effects of the nonlinearities on the systems. Also, it is robust to external changes in the system.

We show some simulation results of the adaptive critic neurocontroller using a two-link robot manipulator. The dynamics equations for an n-link manipulator is given by

$$M(q)\ddot{q} + V(q,\dot{q}) + F(q,\dot{q}) + G(q) + \tau_d = \tau, \tag{7.4.1}$$

where $q(t) \in \mathfrak{R}^n$ are the joint variables, $M(q) \in \mathfrak{R}^{n \times n}$ is the inertia matrix, $V(q,\dot{q}) \in \mathfrak{R}^n$ is the Coriolis vector, $F(q,\dot{q}) \in \mathfrak{R}^n$ is the friction vector, $G(q) \in \mathfrak{R}^n$ is the gravity vector, τ_d is the external disturbances vector, and $\tau \in \mathfrak{R}^n$ is the control torque vector. The above dynamics equation can be written in the Brunovsky form,

$$\begin{aligned} \dot{x}_1 &= x_2, \\ \dot{x}_2 &= f(x) + u + d, \end{aligned} \tag{7.4.2}$$

where $x_1 = [q_1 \quad q_2]^T$, $x_2 = [\dot{q}_1 \quad \dot{q}_2]^T$, $u = M^{-1}(q)\tau$, $d = M^{-1}(q)\tau_d$, and $f(x) = -M^{-1}(q)\{V(q,\dot{q}) + F(q,\dot{q}) + G(q)\}$ is called the nonlinear robot function.

Assuming $M(q)$ is known, $u(t)$ can be computed using Theorem 7.3.1 and then the control torque can be calculated as $\tau = M(q)u$. It is important to remark that nonlinearities such as friction, gravity, and Coriolis terms are unknown.

The actual values of M, V, F, and G for the two-link manipulator used here are given in Lewis, Abdallah, and Dawson (1993).

Here we simulate the trajectory tracking performance of the system for sinusoidal reference signals. The reference signals used for each joint are $q_{d1}(t) = \sin(t)$ radians and $q_{d2}(t) = \cos(t)$ radians.

The simulation parameters are as follows:

- $l_1 = l_2 = 1$m (manipulator arm lengths).

- $m_1 = m_2 = 1$Kg (joint masses).

- Linear control gains: $K_v = \text{diag}[30]$.

- Design parameters: $\Lambda = \begin{bmatrix} 10 & 0 \\ 0 & 7 \end{bmatrix}$, $\Gamma = 25.0$, and $k_z = 3.0$.

- Simulation time: 10 seconds.

The NN architecture for the critic NN is as follows:

- Number of hidden neurons: 10.

- Activation function for hidden neurons: $\sigma(z) = \frac{1}{1+\exp(-z)}$ (sigmoid).

- Input to NN: $[1 \quad r^T]^T$, where $r \in \Re^2$.

- Inputs to hidden neurons: $\chi_1 = V_1^T r$.

The first-layer weights V_1 are set to constant random values in order to provide a basis (Igelnik and Pao (1995)).

The NN architecture for the action-generating NN can be characterized as follows:

- Number of hidden neurons: 10.

- Activation function for hidden neurons: $\sigma(z) = \frac{1}{1+\exp(-z)}$ (sigmoid).

- Input to NN: $x = [1 \quad q^T \quad \dot{q}^T \quad e^T \quad \dot{e}^T \quad r^T \quad q_d^T \quad \dot{q}_d^T]^T$, where $q, \dot{q}, e, \dot{e}, r, q_d, \dot{q}_d \in \Re^2$.

- Inputs to hidden neurons: $\chi_2 = V_2^T x$.

The first-layer weights V_2 are set to constant random values in order to provide a basis.

Case 1: PD gain only. The tracking performance without the output of the action-generating NN is shown in Figure 7.4.1 with $K_v = \text{diag}[30]$. This is the case in Figure 7.3.1, where both NNs are eliminated form the picture. Because of the nonlinearities present in the robot dynamics, the tracking performance is not satisfactory using only the PD loop $K_v r(t)$. A steady-state error results from the nonlinearities in the dynamics.

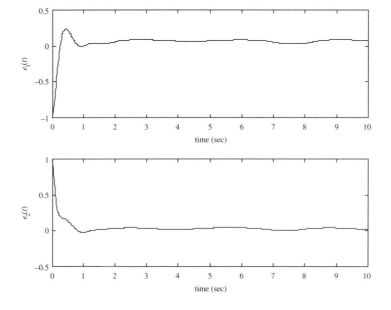

Figure 7.4.1. *Tracking errors with PD control.*

Case 2: PD gain plus adaptive critic neurocontroller. The performance of the adaptive critic neurocontroller is shown in Figure 7.4.2. The tracking error converges to a small value as expected from the stability analysis. This effectively eliminates the steady-state error even though there is no integrator in the feedforward control path. This behavior is due to the memory inherent in NN1 and NN2. The proposed algorithm has good performance in canceling the nonlinearities in the robotic system which are completely unknown to the controller. Only one parameter Γ, the training rate for the action-generating NN, must be selected. This makes the tuning process less complicated than other schemes.

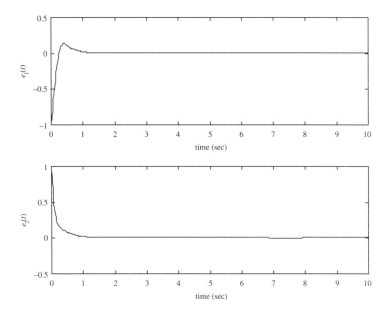

Figure 7.4.2. *Tracking errors with adaptive critic neurocontroller.*

Case 3: PD gain with step mass change. In this case, we change the mass of the second end-effector from 1.0Kg to 2.5Kg at 5 seconds. This corresponds to a payload mass being picked up. Then the end-effector mass is changed back to 1.0Kg at 7 seconds, which corresponds to the payload mass being released. For the first case, we use only the PD loop $K_v r(t)$ with both NNs absent in Figure 7.3.1. Figure 7.4.3 shows that the PD control by itself cannot handle this situation, and the tracking is deteriorated even further when compared with the results in Case 1.

Case 4: PD gain plus adaptive critic neurocontroller with step mass change. In this case, we perform the same change in the mass of the second end-effector as in Case 3 but now using the adaptive critic neurocontroller. The output tracking errors are shown in Figure 7.4.4. After the transient period, the system returns to a small tracking error state, which shows the robustness of the adaptive critic neurocontroller to changes in the system.

7.5   Conclusions

In this chapter, we have presented a method for adaptive critic feedback control of nonlinear systems using feedforward NN. The adaptive critic neurocontroller architecture is simpler

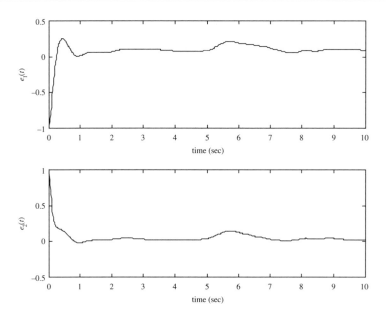

Figure 7.4.3. *Tracking errors for PD control with step mass change.*

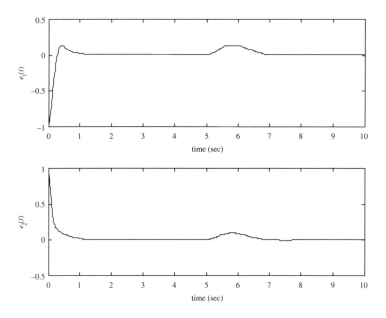

Figure 7.4.4. *Tracking errors for adaptive critic neurocontroller with step mass change.*

than other supervised learning schemes that require gradient information either directly or indirectly. The tuning for the adaptive critic neurocontroller does not require any complicated weight initialization procedure or any preliminary offline training phase; it learns online in real time and offers guaranteed tracking and bounded NN weights and control signals. The adaptive critic neurocontroller is model free in the sense that it works for any system in a prescribed class without the need for extensive modeling and preliminary analysis to find "regression matrices" like those needed in standard adaptive schemes. Unlike standard adaptive schemes, the adaptive critic neurocontroller does not require LIP, persistence of excitation, or CE. The universal approximation property of NNs renders the controller universal and reusable, which means that the same controller will perform even if the behavior or structure of the system has changed.

7.A   Appendix: Proof of Theorem 7.3.1

As stated in (7.3.9), the system dynamics are

$$\dot{r} = -K_v r + \tilde{W}_2^T \sigma(\chi_2) + \varepsilon(x) + d(t) + v(t). \tag{7.A.1}$$

Define the matrices

$$\hat{W} = \begin{bmatrix} \hat{W}_1 & 0 \\ 0 & \hat{W}_2 \end{bmatrix} \quad \text{and} \quad \alpha = \begin{bmatrix} r \\ R \end{bmatrix}. \tag{7.A.2}$$

Define the Lyapunov function candidate

$$L = \frac{1}{2}\alpha^T \alpha + \frac{1}{2}\operatorname{tr}(\tilde{W}^T \Gamma^{-1} \tilde{W}). \tag{7.A.3}$$

Differentiating and using (7.A.2),

$$\dot{L} = \alpha^T \dot{\alpha} + \operatorname{tr}(\tilde{W}^T \Gamma^{-1} \dot{\tilde{W}}) = r^T \dot{r} + R^T \dot{R} + \operatorname{tr}(\tilde{W}^T \Gamma^{-1} \dot{\tilde{W}}). \tag{7.A.4}$$

Substituting (7.3.9), (7.3.11), and (7.A.2) in (7.A.4) yields

$$\dot{L} = r^T [-K_v r + \tilde{W}_2^T \sigma(\chi_2) + \varepsilon(x) + d(t) + v(t)]$$
$$+ R^T [\dot{\hat{W}}_1^T \sigma(\chi_1) + \hat{W}_1^T \sigma'(\chi_1)\dot{\chi}_1 + \dot{\rho}] + \operatorname{tr}(\tilde{W}^T \Gamma^{-1} \dot{\tilde{W}})$$
$$= -r^T K_v r + r^T [\varepsilon(x) + d(t) + v(t)]$$
$$+ \operatorname{tr}(-\tilde{W}_1^T \Gamma_1^{-1} \dot{\hat{W}}_1 - \tilde{W}_2^T \Gamma_2^{-1} \dot{\hat{W}}_2 + \tilde{W}_2^T \sigma(\chi_2)r^T + \dot{\hat{W}}_1^T \sigma(\chi_1)R^T$$
$$+ \hat{W}_1^T \sigma'(\chi_1)\dot{\chi}_1 R^T) + R^T \dot{\rho}.$$

Using the fact that $\dot{\chi}_1 = V_1^T \dot{r}$ and the dynamics in (7.3.9) for the second time,

$$\dot{L} = -r^T K_v r + r^T [\varepsilon(x) + d(t) + v(t)] + R^T \dot{\rho}$$
$$+ \operatorname{tr}(\tilde{W}_2^T [\sigma(\chi_2)r^T - \Gamma_2^{-1} \dot{\hat{W}}_2] + \dot{\hat{W}}_1^T [\sigma(\chi_1)R^T - \Gamma_1^{-1} \dot{\hat{W}}_1]$$
$$+ \hat{W}_1^T \sigma'(\chi_1)V_1^T [-K_v r + \tilde{W}_2^T \sigma(\chi_2) + \varepsilon(x) + d(t) + v(t)]R^T).$$

Using the property $\operatorname{tr}(AB) = \operatorname{tr}(BA)$, one has

$$\dot{L} = -r^T K_v r + r^T [\varepsilon(x) + d(t) + v(t)]$$
$$+ \operatorname{tr}(\tilde{W}_2^T [\sigma(\chi_2)r^T - \Gamma_2^{-1} \dot{\hat{W}}_2 + \sigma(\chi_2)R^T \hat{W}_1^T \sigma'(\chi_1)V_1^T])$$
$$+ \operatorname{tr}(\dot{\hat{W}}_1^T [\sigma(\chi_1)R^T - \Gamma_1^{-1} \dot{\hat{W}}_1]$$
$$+ \hat{W}_1^T \sigma'(\chi_1)V_1^T [-K_v r + \varepsilon(x) + d(t) + v(t)]R^T) + R^T \dot{\rho}.$$

Substituting (7.3.12) and (7.3.13) in the previous equation yields

$$
\begin{aligned}
\dot{L} = {}& -r^T K_v r + r^T [\varepsilon(x) + d(t) + v(t)] + \operatorname{tr}(\tilde{W}_2^T \hat{W}_2) + R^T \dot{\rho} \\
& + \operatorname{tr}(-R\sigma(\chi_1)^T \sigma(\chi_1) R^T - R\sigma(\chi_1)^T \Gamma_1 \hat{W}_1 + \tilde{W}_1^T \Gamma_1^{-1} \sigma(\chi_1) R^T) \\
& + (\tilde{W}_1^T \hat{W}_1 \hat{W}_1^T \sigma'(\chi_1) V_1^T [-K_v r + \varepsilon(x) + d(t) + v(t)] R^T) \\
= {}& -r^T K_v r + [(V_1 \sigma'(\chi_1)^T \hat{W}_1 R) + r]^T [\varepsilon(x) + d(t) + v(t)] + \operatorname{tr}(\tilde{W}^T \hat{W}) + R^T \dot{\rho} \\
& - R^T \sigma(\chi_1)^T \sigma(\chi_1) R - R^T \Gamma_1 \hat{W}_1^T \sigma(\chi_1) + R^T \tilde{W}_1^T \Gamma_1^{-1} \sigma(\chi_1) - R^T \hat{W}_1^T \sigma'(\chi_1) V_1^T K_v r \\
= {}& -r^T K_v r + [(V_1 \sigma'(\chi_1)^T \hat{W}_1 R) + r]^T [\varepsilon(x) + d(t) + v(t)] \\
& + \operatorname{tr}(\tilde{W}^T (W - \tilde{W})) + R^T \dot{\rho} - R^T \sigma(\chi_1)^T \sigma(\chi_1) R - R^T \Gamma \hat{W}_1^T \sigma(\chi_1) \\
& + R^T W_1^T \Gamma_1^{-1} \sigma(\chi_1) - R^T \hat{W}_1^T \Gamma_1^{-1} \sigma(\chi_1) - R^T \hat{W}_1^T \sigma'(\chi_1) V_1^T K_v r.
\end{aligned}
$$

Finally, using (7.3.14), we can get a bound on $\dot{L}$ as

$$
\begin{aligned}
\dot{L} \leq {}& -K_{\min}\|r\|^2 + [(V_1 \sigma'(\chi_1)^T \hat{W}_1 R) + r]^T [\varepsilon(x) + d(t) + v(t)] \\
& + W_{\max}\|W^\sim\| - \|W^\sim\|^2 - \|\sigma(\chi_1)\|^2 \|R\|^2 + \|R\| \cdot \|W_1\| \cdot \|\sigma(\chi_1)\|.
\end{aligned}
\tag{7.A.5}
$$

Completing the squares for $\|R\|$ and $\|W^\sim\|$,

$$
\begin{aligned}
\dot{L} \leq {}& -K_{\min}\|r\|^2 + [(V_1 \sigma'(\chi_1)^T \hat{W}_1 R) + r]^T [\varepsilon(x) + d(t) + v(t)] \\
& - \left[\|\tilde{W}\| - \frac{W_{\max}}{2}\right]^2 + \frac{W_{\max}^2}{4} - \|\sigma(\chi_1)\|^2 \left[\|R\| - \frac{W_{\max}}{2\|\sigma(\chi_1)\|}\right]^2 + \frac{W_{\max}^2}{4}.
\end{aligned}
\tag{7.A.6}
$$

Using the robustifying term defined in (7.3.10), we get that $L \leq 0$ as long as

$$
\|r\| \geq \frac{W_{\max}^2}{\sqrt{2K_{v\min}}} \quad \text{and} \quad \|\tilde{W}\| \geq W_{\max} \frac{1 + \sqrt{2}}{2}.
\tag{7.A.7}
$$

From the equations in (7.A.7), $\dot{L}$ is negative outside a compact set. According to a standard Lyapunov theorem extension, it can be concluded that the tracking error $r(t)$ and the NN weight estimates $\tilde{W}_1$ and $\tilde{W}_2$ are GUUB. $\qquad \square$

Chapter 8

Neurocontrol of Telerobotic Systems with Time Delays

Guest author: Jinquan Huang[1]

This chapter develops a new recurrent NN predictive control (RNNPC) strategy to compensate for input and feedback time delays in telerobotic systems with communication time delays. The proposed control structure consists of a local linearized subsystem and a remote predictive controller. In the local linearized subsystem, a recurrent NN (RNN) with an online weight-tuning algorithm is employed to approximate the dynamics of the time-delay-free nonlinear plant. The remote controller is a modified Smith predictor for the local linearized subsystem that provides prediction and maintains the desirable tracking performance in the presence of time delays. Stability analysis is given in the sense of Lyapunov. The result is an adaptive compensation scheme for unknown telerobotic systems with time delays, uncertainties, and external disturbances. A simulation of a two-link robotic manipulator is provided to illustrate the effectiveness of the proposed control strategy.

8.1   Introduction

The human's capacity to perform certain tasks is limited by a lack of strength and an inability to adapt to and perform effectively in severe environments. One example is future assembly work in outer space, where the environment is inaccessible to humans because of harsh conditions and such work cannot be directly performed by humans. A solution to this problem is found by using telerobotic systems.

   In telerobotic systems, a human operator conducts a task, moving a master robot controller and thus defining motion and force commands. The master is connected through a communication channel to a slave robot manipulator in a remote location whose purpose is to mimic the master, thus performing the commanded motions and exerting the commanded forces on its environment. A typical telerobotic system is depicted in Figure 8.1.1. Task performance is enhanced if the human operator has information on the contact force being exerted by the slave manipulator. A convenient way of providing this information is to "reflect" the contact force to the motors in the master controller, so the operator can feel a resistive force indicative of the contact force.

   Unfortunately, if force feedback is used, the presence of time delay, mainly caused by the communication channels, may destabilize the telerobotic control system and degrade

---

[1]Nanjing University of Aeronautics and Astronautics, Nanjing, People's Republic of China. This research was performed while the author was a Visiting Researcher at the University of Texas at Arlington.

Figure 8.1.1. *Typical telerobotic system.*

the performance of the remotely controlled manipulator. This time delay may be in the seconds or minutes range. Since the operator cannot in a timely fashion receive the correct information about the current motion of the remote robot and the force interaction between the manipulator and the environment, the correct control commands cannot be issued by the operator. Therefore, successful teleoperation cannot be expected without conducting effective compensation for the time delay.

Much effort has been devoted to solving this problem (Sheridan (1993)). Anderson and Spong (1989) proposed a bilateral master–slave system in which the operator sent commands to the slave, and the slave in turn sent velocity information back to the operator. In their control scheme, they introduced passivity and a scattering operator and proved that using a scattering matrix representation, the stability in passively controlled systems can be ascertained. Using the passivity approach provides an easy method for guaranteeing stability of a system, yet the actual system performance is not quantified and may or may not be appropriate for the tasks at hand. This is particularly true in the case of time-delayed teleoperation, where the system behavior is quite complex and should degrade smoothly as the delay increases.

Hannaford (1989) applied a hybrid two-port model to teleoperators with force and velocity sensing at the master and slave. To deal with the time delay, he proposed a teleoperator architecture that was governed by a bilateral impedance control. Depending on the existence of very complex estimators capable of identifying the impedance of the environment and of the human operator, this bilateral impedance control exhibited stability in the presence of time delay. Hannaford and Kim (1989) quantified the performance of a telerobotic system with respect to time delays under two modes of system control: kinesthetic force feedback and shared compliance control. Tests were conducted to observe the effectiveness of the teleoperator control in the presence of time delays. It was determined that task performance was not possible using the kinesthetic force feedback control for delays greater than 1 second. The shared compliance control resulted in good performance for delays up to 4 seconds.

Sheridan (1993) discusses a kinematic predictor display that used a geometric model of the manipulator to superpose the commanded positions over the image of the delayed actual video. Buzan and Sheridan (1989) used this concept in their predictive operator aid, which provided position and force predictions to the operator. This was implemented with the use of a dual visual feedback predictive display and dual predictive force reflectors. The performance of these techniques depended on the model and calibration errors and on having relatively fixed environmental objects.

A study of time-delay systems is given by Malek-Zavarei and Jamshidi (1987). Robots are nonlinear systems. Many kinds of time-delay compensation methods have been presented for linear systems. One of the earliest controllers for constant time-delay systems, still relevant today, is the *Smith predictor* (Smith (1957)). Attention has been paid to this scheme over the years and some of its properties have been reported by Alevisakis and Seborg (1973), Laughlin, Rivera, and Morari (1987), Mee (1971), and Palmor (1980). According to Smith's principle, a predictor is equivalent to the time-delay-free part of the plant. Then a controller exerts control according to the feedback information from the predictor. However, the Smith prediction controller requires a good model of the plant. In the face of inevitable mismatch between the model and the actual plant, the closed-loop performance may be very poor. Moreover, if the plant is nonlinear or there exists a disturbance during the control process, it is difficult to use a linear model–based control strategy to obtain satisfactory control performance. For highly nonlinear processes, time-delay compensation techniques that employ nonlinear models can significantly improve system performance. However, few nonlinear time-delay compensation strategies are currently available (Bequette (1991), Henson and Seborg (1994), Kravaris and Wright (1989), Lee and Sullivan (1990)). The Smith predictor was applied to known telerobotic systems with fairly good results by Lee and Lee (1994).

As discussed in previous chapters, NNs have been widely applied to identification and control for nonlinear systems. Theoretical work has proven that NNs can uniformly approximate any continuous function over a compact domain, provided the network has a sufficient number of neurons. Recently, interest has been increasing towards the usage of dynamical NNs for modeling and identification of dynamical systems (Cheng, Karjala, and Himmelblau (1995), Karakasoglu, Sudharsanan, and Sundareshan (1993), Kosmatopoulos et al. (1995)). These networks, which naturally involve dynamic elements in the form of feedback connections, are known as dynamic RNNs.

In this chapter, we consider control of a telerobotic system with constant time delays. An NN-based controller is designed for this class of telerobotic systems with constant input and state-feedback delays arising from a communication channel. An RNN is used to identify the delay-free nonlinear robot dynamical system. The RNN weights are tuned online, with no preliminary offline "learning phase" needed. The nonlinear term is canceled using a local nonlinear compensation based on the RNN output. The time-delay effect is canceled using a modified Smith predictive method with an extra robustifying term. One of the difficulties encountered in nonlinear time-delay control systems is the guaranteed stability of the overall system. In this chapter, stability properties of the closed-loop system are guaranteed using Lyapunov stability theory. The result is an adaptive time-delay compensation scheme for nonlinear telerobotic systems.

8.2   Background and problem statement

The dynamics of an n-link robot manipulator may be expressed in the Lagrange form (Lewis, Abdallah, and Dawson (1993))

$$M(q)\ddot{q} + V(q,\dot{q}) + G(q) + F(\dot{q}) + \tau_d = \tau, \qquad (8.2.1)$$

with $q(t) \in \Re^n$ the joint variable vector, $M(q)$ the inertia matrix, $V(q,\dot{q})$ the Coriolis/centripetal vector, $G(q)$ the gravity vector, and $F(\dot{q})$ the friction vector. Bounded unknown disturbances (including, e.g., unstructured unmodelled dynamics) are denoted by $\tau_d(t)$, and the control input torque vector is $\tau(t)$.

In this work, we assume that there is a communications link that imposes a constant delay T in the input channel and the output channel. That is, the joint variable $q(t)$ can only be measured after a delay of T and the control input $\tau(t)$ can only be applied to the system after a delay of T. An NN and a modified Smith predictor will be used to perform output tracking of the robotic system (8.2.1) with time delays. The control objective can be described as follows: given a desired arm trajectory $q_d(t)$, find a control $\tau(t)$ such that the robot follows the delayed desired trajectory with an acceptable accuracy (i.e., bounded error tracking), while all the states and controls remain bounded.

All rigid-link robot arm dynamics satisfy the following properties required in the analysis.

Property 1. $M(q)$ is a positive definite symmetric matrix bounded by

$$m_1 I \leq M(q) \leq m_2 I \tag{8.2.2}$$

with m_1 and m_2 known positive constants.

Property 2.   The unknown disturbance vector $\tau_d(t)$ satisfies $\|\tau_d(t)\| \leq b_d$ with b_d a known positive constant.

In practical robotic systems, uncertainties that may affect the tracking performance are inevitable. Hence it is important to consider the effects due to uncertainties. In this chapter, robot uncertainties and external disturbances will be considered simultaneously. An RNN is introduced in the next section to estimate these unknown or uncertain dynamics by an online tuning algorithm.

In order to estimate the uncertainties, it is necessary to derive a suitable state-space dynamic equation. The robot dynamics equation (8.2.1) can be expressed in the form of a state equation with the state vector $x = \begin{pmatrix} q \\ \dot{q} \end{pmatrix}$ as

$$\dot{x} = \begin{bmatrix} [0_{n \times n} \quad I_{n \times n}]x \\ -M^{-1}(x)[V(x) + G(x) + F(x)] \end{bmatrix} + \begin{bmatrix} 0_{n \times n} \\ M^{-1}(x) \end{bmatrix} \tau - \begin{bmatrix} 0_{n \times n} \\ M^{-1}(x) \end{bmatrix} \tau_d, \tag{8.2.3}$$

where $I_{n \times n}$ and $0_{n \times n}$ denote the $(n \times n)$-dimensional identity matrix and zero matrix, respectively.

We assume that the inertia matrix $M(x)$ is known. It is usual to have uncertainties in the Coriolis terms, which are difficult to compute, and the friction term, which may have a complicated form. Hence (8.2.3) can be rearranged into the form

$$\dot{x} = A_S x + B\{f(x) + M^{-1}(x)\tau\} - BM^{-1}(x)\tau_d, \tag{8.2.4}$$

where

$$A_S = \begin{bmatrix} 0_{n \times n} & I_{n \times n} \\ A_{21} & A_{22} \end{bmatrix}, \qquad B = \begin{bmatrix} 0_{n \times n} \\ I_{n \times n} \end{bmatrix},$$

and $f(x)$ consists of all the unknown terms in the plant, expressed as

$$f(x) = -M^{-1}(x)\{V(x) + G(x) + F(x)\} - [A_{21} \quad A_{22}]x. \tag{8.2.5}$$

We assume that A_{21} and A_{22}, which are $(n \times n)$-dimensional matrices, are suitably chosen such that A_S is stable; i.e., for some symmetric positive definite matrix Q, there exists a symmetric positive definite matrix P satisfying the Lyapunov equation

$$A_S^T P + P A_S = -Q. \tag{8.2.6}$$

Now define a control input torque as

$$\tau = -M(x)\hat{f}(x) + M(x)u \tag{8.2.7}$$

with $\hat{f}(x)$ an estimate of $f(x)$ and $u(t)$ a new control variable. The closed-loop robotic system becomes

$$\dot{x} = A_S x + Bu + B\tilde{f}(x) - BM^{-1}(x)\tau_d, \tag{8.2.8}$$

where the functional estimation error is given by

$$\tilde{f} = f - \hat{f}. \tag{8.2.9}$$

If the parameter matrices $M(x)$, $V(x)$, $G(x)$, and $F(x)$ of robotic systems are well known and available in the design of control law, the term $f(x)$ in (8.2.4) can be exactly canceled by a suitable choice of control law $\tau(t)$, and the system is linearized by using control input torque (8.2.7) in (8.2.4). Then we can use the Smith predictor method to cancel the time-delay effect for telerobotic systems. However, in practical robotic systems, parametric uncertainties that cannot be known exactly a priori are inevitable. Unfortunately, linear parametrization is usually used in the derivation of existing standard adaptive control results (see Ortega and Spong (1989) and references therein), and the system parameters are assumed to be constant or slowly varying in analyzing the adaptive control design. However, in practice, the system parameters may be quickly varying or the linear parametrizable property may not hold. Moreover, computation of the regressor matrix required in most adaptive control approaches is a time-consuming task. On the other hand, besides parametric uncertainties, a robotic system may also be perturbed by unmodelled dynamics and external disturbances. Hence the introduction of an alternative and suitable method for learning behavior of the total dynamic system is of interest.

In this chapter, an RNN estimation approach will be proposed to solve this problem. For the control problem, an NN and a Smith predictor will be used to perform output tracking of the system (8.2.1) with time delays in the input channel and the output channel. Our objective can be described as follows: Given a desired trajectory $q(t) \in \Re^n$, find a control $u(t)$ such that the telerobotic manipulator follows the delayed desired trajectory with an acceptable accuracy (i.e., bounded error tracking), while all the states and controls remain bounded.

First, define a desired trajectory vector

$$x_d(t) = [q_d^T(t) \quad \dot{q}_d^T(t)]^T. \tag{8.2.10}$$

We shall make some mild assumptions that are widely used.

Assumption 1. The desired trajectory vector $x_d(t)$ and its derivative $\dot{x}_d(t)$ are continuous and available for measurement, and $x_d(t)$ and $\dot{x}_d(t)$ are bounded.

Assumption 2. The delay time T is a known constant, and the delay time in the input channel is equal to that in the feedback channel.

Assumption 3. $x(t)$ is assumed to be completely measurable and available for the local controller. The delayed value $x(t - T)$ is available to the remote human operator controller.

8.3 Identification of time-delay-free systems using recurrent neural networks

The design of the NN control system for the telerobotic system with delay proceeds in two steps. First we design an NN estimator for the undelayed robot system. Then we design a modified Smith predictor that compensates for the system delay. The NN estimator can be viewed as belonging to a "local" controller, while the Smith predictor can be viewed as belonging to a "remote" controller at the other end of the communication link, where the human operator resides (see Figure 8.4.4).

In order to estimate the unknown part (8.2.5) of the time-delay-free dynamic system (8.2.4), we employ an RNN. RNN models are characterized by a two-way connectivity between neuronal units. This distinguishes them from feedforward NNs, where the output of one unit is connected only to units in the next layer. The RNN is described by the set of differential equations

$$\dot{\hat{x}} = A_S \hat{x} + B \hat{W}^T \phi(x) + B M^{-1}(x)\tau, \tag{8.3.1}$$

where $\hat{x} \in \Re^{2n}$ is a state vector, $\hat{W} \in \Re^{m \times n}$ is a synaptic weight matrix, and $\phi(x) \in \Re^m$ is a basis function vector with elements $\phi_i(x)$ smooth, monotone increasing functions. Typical selections for $\phi_i(x)$ include, with $y \in \Re$,

$$\phi_i(y) = \frac{1}{1 + e^{-\alpha y}}, \qquad \text{sigmoid,}$$

$$\phi_i(y) = \frac{e^{\alpha y} - e^{-\alpha y}}{e^{\alpha y} + e^{-\alpha y}}, \qquad \text{hyperbolic tangent (tanh),} \tag{8.3.2}$$

$$\phi_i(y) = e^{-\frac{(y - m_i)^2}{s_i}}, \qquad \text{radial basis functions (RBF).}$$

Since we employ a one-layer NN here, it is important that $\phi(x)$ be selected as a basis (Chapter 1).

Due to the approximation capabilities of the RNNs, we can assume without loss of generality that the unknown nonlinear function $f(x)$ in system (8.2.4) can be completely described on a compact set S by an RNN plus a bounded reconstruction error ε so that $f(x) = W^T \phi(x) + \varepsilon$ when $x \in S$ for some ideal weights W, generally unknown. Then system (8.2.4) can be written on a compact set S as

$$\dot{x} = A_S x + B W^T \phi(x) + B M^{-1}(x)\tau - B M^{-1}(x)\tau_d + \varepsilon. \tag{8.3.3}$$

Standard assumptions, quite common in the NN literature, are stated next.

Assumption 4. The RNN reconstruction error is bounded so that $\|\varepsilon\| < \varepsilon_N$ with the bound ε_N a constant on a compact set S.

Assumption 5. The ideal weights are unknown but bounded by a known positive value W_M so that $\|W\|_F \le W_M$.

Assumption 6. The input $\tau(t)$ and the state $x(t)$ remain bounded for all $t \ge 0$.

This last assumption is needed only in this section for identification purposes. In the subsequent section on control, this assumption is dropped since it will be necessary to prove that the proposed controller keeps both $x(t)$ and $\tau(t)$ bounded.

In order to provide robustness to the unknown ε and disturbance τ_d, we add a robustifying term $v(t)$ in (8.3.1) so that the NN estimator becomes

$$\dot{\hat{x}} = A_S \hat{x} + B \hat{W}^T \phi(x) + B M^{-1}(x)\tau + v. \tag{8.3.4}$$

Subtracting (8.3.4) from (8.3.3) yields the error equation

$$\dot{\tilde{x}} = A_S \tilde{x} + B \tilde{W}^T \phi(x) - B M^{-1}(x)\tau_d + \varepsilon - v, \tag{8.3.5}$$

where $\tilde{x} = x - \hat{x}$ is the *estimation error* between the real state and the estimator state and $\tilde{W} = W - \hat{W}$ is the weight estimation error between the unknown ideal weights and the actual weights in the NN.

The next result is our main identification result. It is used later to derive a controller. It shows how to tune the RNN weights to obtain an estimate of the unknown dynamics. UUB is defined in Chapter 2.

Theorem 8.3.1 (System identification using RNNs). *Given Assumptions 3–6, consider the robot manipulator system (8.2.1) or (8.3.3) and the RNN estimator given by (8.3.4). Let the RNN weight-tuning algorithm be provided by*

$$\dot{\hat{W}} = F\phi(x)\tilde{x}^T P B - K_e F\|\tilde{x}\|\hat{W} \qquad (8.3.6)$$

with any constant matrix $F = F^T > 0$ and scalar design parameter $K_e > 0$. Let the robustifying term be provided by

$$v = K_r \tilde{x} + K_W \|\hat{W}\|_F \tilde{x} \qquad (8.3.7)$$

with scale constants $K_r > 0$ and $K_W > 0$. Then the estimation error $\tilde{x}(t)$ and the NN weight estimate $\hat{W}(t)$ are UUB with specific bounds given by

$$\|\tilde{x}\| > \frac{P_M\left(\sqrt{n}m_1^{-1}b_d + \varepsilon_N\right) + \frac{1}{4}K_e W_M^2}{\frac{1}{2}Q_m + K_r P_m}, \qquad (8.3.8)$$

$$\|\tilde{W}\|_F > \frac{1}{2}W_M + \sqrt{\frac{P_M\left(\sqrt{n}m_1^{-1}b_d + \varepsilon_N\right)}{K_e} + \frac{1}{4}W_M^2}, \qquad (8.3.9)$$

and $\|\tilde{x}\|$ may be made as small as desired by increasing the gain K_r.

Proof. See the first appendix (section 8.A) of this chapter.

For practical purposes, (8.3.8) and (8.3.9) can be considered as bounds on $R(s)$ and $\|\tilde{W}\|_F$. Note that the bound on the approximation error may be kept arbitrarily small if the gain K_r in the robustifying term is chosen big enough. On the other hand, (8.3.9) reveals that the RNN weight errors are fundamentally bounded by W_M. The parameter $G_z(s)$ offers a design trade-off between the relative eventual magnitudes of $\|\tilde{x}\|$ and $\|\tilde{W}\|_F$.

The first term of (8.3.6) is nothing but a continuous-time version of the standard backpropagation algorithm. The second term in (8.3.6) corresponds to the *e-mod* (Narendra and Annaswamy (1987)) in standard use in adaptive control to guarantee bounded parameter estimates. Thus the weight-tuning algorithm (8.3.6) does not require the *persistence of excitation* condition.

The identifier convergence is ensured only if the input $\tau(t)$ and state $x(t)$ are assumed bounded (Assumption 6). This assumption is quite restrictive. However, this assumption will be eliminated in the section on closed-loop feedback control design, where the identifier is combined with a Smith predictor in the closed-loop system for trajectory control.

8.4 Recurrent neural network predictive control strategy for compensating time delays

The Smith predictor can be used to control known linear systems with constant time delays. Unfortunately, robotic systems are neither linear nor known. In this section, we propose a control structure that extends the Smith predictor technique (Smith (1957)) to unknown nonlinear robotic systems with disturbances. The key is to use an NN local controller to compensate for nonlinearities and disturbances and approximately linearize the robot dynamics. First, we review the standard Smith predictor and put it into a more useful form.

8.4.1   Review of the Smith predictor

Consider a conventional time-delay linear control system shown in Figure 8.4.1, where $G_p(s)$ is the plant, $G_c(s)$ is the controller, $R(s)$ and $C(s)$ are the input and output of the system, respectively, and T is the known delay time. If the fictitious variable $Y(s)$ in Figure 8.4.1 were measurable and fed back to the cascade controller, then it would move the time delay outside the control loop. Under this condition, the system output is simply the delayed value of the delay-free portion of the system. However, this separation of the delay from the system is not possible in most cases.

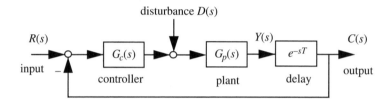

Figure 8.4.1. *A typical time-delay system.*

In order to improve the design of the time-delay system, one can model the plant as indicated by Figure 8.4.2, where $G_z(s)$ represents the dynamic model of the plant, called the *predictor*. Although the fictitious variable $Y(s)$ is not available, $Z(s)$ can be used for feedback purposes as shown in Figure 8.4.2. This scheme would only control the model well but not the system output $C(s)$. Inaccurate models and disturbances cannot be accommodated. In order to compensate for these errors, a second feedback loop is introduced as shown by dashed lines in Figure 8.4.2. This is the Smith predictor control scheme for delay systems. In this way, the effect of the time delay in the feedback loop is minimized, allowing one to use conventional controllers such as PD, PI, or PID for G_c.

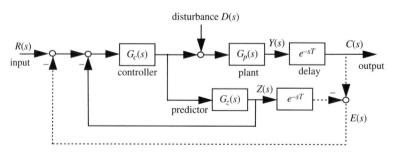

Figure 8.4.2. *Classical Smith predictor control system.*

An alternative representation of the Smith predictor that is more useful for our purposes is shown in Figure 8.4.3. Assuming that disturbance $D(s) = 0$, the closed-loop transfer function of the system is given by

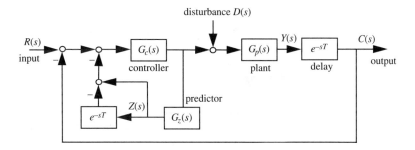

Figure 8.4.3. *An alternative schematic for the Smith predictor.*

$$\frac{C(s)}{R(s)} = \frac{G_c(s)G_p(s)e^{-sT}}{1 + G_c(s)G_z(s) - G_c(s)G_z(s)e^{-sT} + G_c(s)G_p(s)e^{-sT}}. \tag{8.4.1}$$

If $G_p(s) = G_z(s)$, the above equation is reduced to

$$\frac{C(s)}{R(s)} = \frac{G_c(s)G_p(s)e^{-sT}}{1 + G_c(s)G_z(s)}. \tag{8.4.2}$$

As can be seen from the above relation, the effect of the delay has been eliminated from the denominator of the transfer function. The system output is the delayed value of the delay-free portion of the system.

8.4.2 Control structure

Figure 8.4.4 shows the structure of the proposed RNNPC strategy for telerobotic systems with time delays. This system is composed of three parts:

- remote predictive controller—at the human operator;

- communication channel;

- local linearized subsystem—at the robotic system.

The *communication channel* has two time delays. One is in the input channel and the other is in the feedback channel.

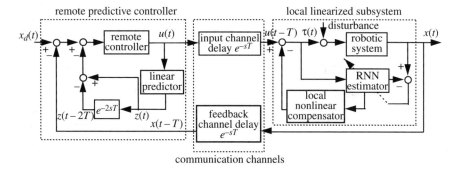

Figure 8.4.4. *RNNPC strategy for telerobotic system with time delay.*

In this case, it might be useful to use the Smith predictor principle discussed in the previous subsection to deal with the time delays. Since the Smith predictor is usually based on a linear model, we must extend its concept to the nonlinear case. First, in the so-called *local linearized subsystem*, we use an estimator to identify the dynamics of the unknown nonlinear robot. If the estimator can accurately approximate the true plant, we can use a local feedback, called the local nonlinear compensator, to cancel the nonlinear effects. Thus the local subsystem is linearized and a stable linear predictor model is obtained at the same time.

The *remote predictive controller* can deal with the time delay for the local linearized subsystem based on the principle of the Smith predictor. If the linear predictor is an accurate model of the local linearized subsystem, then referring to Figure 8.4.4, $z(t - 2T) = x(t - T)$ and $z(t) = x(t + T)$. This means that the output of the controlled plant $x(t)$ ahead of time T can be predicted via the predictor's output $z(t)$. Therefore, we overcome the difficulty in controlling telerobots with time delays.

In the next section, we will use an RNN model to approximate the robot dynamics. As will be seen, the linear part of the RNN model is known and invariant. The RNN weights are tuned online so that the proposed controller has good modeling error and disturbance rejection properties.

8.5 Recurrent neural network predictive control and stability analysis

The goal in this section is to cancel the nonlinear terms of the robot dynamics using a local nonlinear feedback, in order to have approximately linear dynamics between a new input and the actual plant output. We call this *local nonlinear compensation*. Then we can use the Smith predictor principle to design the remote predictive control. A rigorous proof of the stability of this scheme is offered. The overall RNNPC structure for the telerobotic system with time delay is shown in Figure 8.5.1.

According to section 8.3, the unknown system can be modeled by an RNN:

$$\dot{x}(t) = A_S x(t) + BW^T \phi(x(t)) + BM^{-1}(x(t))\tau(t) - M^{-1}(x(t))\tau_d(t) + \varepsilon(t). \quad (8.5.1)$$

The estimator (8.3.4) for this model can be obtained using the online weight-tuning law

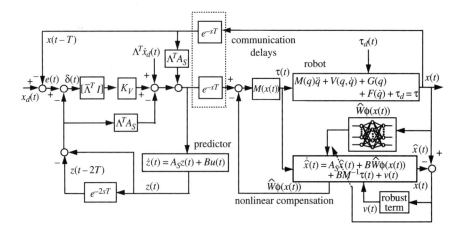

Figure 8.5.1. *Overall RNNPC structure for telerobotic system with time delay.*

(8.3.6) and the robustifying term (8.3.7). To locally compensate the nonlinear part of the system, let the local control input be given by

$$\tau(t) = M(x(t))u(t - T) - M(x(t))\hat{W}^T(t)\phi(x(t)), \qquad (8.5.2)$$

where $u(t - T)$ stands for a new control variable from the output of the remote controller with time delay T in the input channel, which will be designed later. Then the local system becomes a linear model plus a disturbance

$$\dot{x}(t) = A_S x(t) + Bu(t - T) + B\tilde{W}^T(t)\phi(x(t)) - BM^{-1}(x(t))\tau_d(t) + \varepsilon(t). \qquad (8.5.3)$$

For the local linearized system (8.5.3) predictive control will be used to perform output tracking. In the following, we will find a control action $u(t)$ such that the robot state $x(t)$ follows the delayed desired trajectory $x_d(t - T)$ with an acceptable accuracy.

Now let the linear predictor be

$$\dot{z}(t) = A_S z(t) + Bu(t). \qquad (8.5.4)$$

We can use the state of the predictor $z(t)$ as a state feedback to design the remote controller. However, due to the possible presence of disturbances and modeling errors, we use $x(t - T) + z(t) - z(t - 2T)$ as the feedback signal to incorporate the Smith predictor loop in Figure 8.4.4.

Define an error vector

$$\delta(t) = x_d(t) - x(t - T) - z(t) + z(t - 2T). \qquad (8.5.5)$$

Define a filtered error as

$$r(t) = \Lambda^T \delta(t) = [\bar{\Lambda}^T \quad I_{n \times n}]\delta(t), \qquad (8.5.6)$$

where $\bar{\Lambda} = \bar{\Lambda}^T > 0$ is a design parameter matrix, usually selected diagonal. Differentiating $r(t)$ and using (8.5.3) and (8.5.4), the filtered error dynamics may be written as

$$\begin{aligned}
\dot{r}(t) &= \Lambda^T \dot{\delta}(t) \\
&= \Lambda^T[\dot{x}_d(t) - \dot{x}(t - T) - \dot{z}(t) + \dot{z}(t - 2T)] \qquad (8.5.7) \\
&= \Lambda^T\{\dot{x}_d(t) - A_S[x(t - T) + z(t) - z(t - 2T)]\} - u(t) \\
&\quad - \Lambda^T[B\tilde{W}^T(t - T)\phi(x(t - T)) + BM^{-1}(x(t - T))\tau_d(t - T) + \varepsilon(t - T)].
\end{aligned}$$

Now define a control input as

$$u(t) = \Lambda^T\{\dot{x}_d(t) - A_S[x(t - T) + z(t) - z(t - 2T)]\} + K_V r(t). \qquad (8.5.8)$$

Then (8.5.7) becomes

$$\begin{aligned}
\dot{r}(t) = -K_V r(t) - \Lambda^T[B\tilde{W}^T(t - T)\phi(x(t - T)) \qquad (8.5.9) \\
+ BM^{-1}(x(t - T))\tau_d(t - T) + \varepsilon(t - T)].
\end{aligned}$$

Due to the presence of the time delay, the weight-tuning algorithm given by (8.3.6) must be modified as follows:

$$\dot{\hat{W}} = F\phi(x)\tilde{x}^T PB - K_e F\|\tilde{x}\|\hat{W} - K_\sigma F\hat{W} \qquad (8.5.10)$$

with scalar design parameter $K_\sigma > 0$. The robustifying term is provided by (8.3.7) as

$$v = K_r \tilde{x} + K_W \|\hat{W}\|_F \tilde{x}. \tag{8.5.11}$$

Assume that the parameter K_σ is chosen such that

$$K_\sigma > \frac{1}{32} \Lambda_M \gamma \sqrt{n}, \tag{8.5.12}$$

where γ is the bound on the value of $\phi(.)$. Then we obtain the following result, which is our main result of the chapter and extends the Smith predictor to nonlinear robotic systems.

Theorem 8.5.1 (RNN controller for time-delay telerobot). *Let Assumptions 1–5 hold. Under the condition (8.5.12), consider the robot manipulator (8.2.1) with input and feedback time delays. Take the control input for the telerobotic system as (8.5.2) and (8.5.8). Let the RNN weight-tuning algorithm be provided by (8.5.10) and the robustifying term be provided by (8.5.11). Then for a large enough control gain K_V, the filtered error $r(t)$, the estimation error $\tilde{x}(t)$, and the weight-tuning error $\tilde{W}(t)$ are UUB with practical bounds given specifically by*

$$\|r(t)\| > \frac{\Lambda_M \left(\sqrt{n} m_1^{-1} b_d + \varepsilon_N \right)}{K_V + 8\Lambda_M \gamma \sqrt{n}},$$

$$\|\tilde{x}(.)\| > \frac{P_M \left(\sqrt{n} m_1^{-1} b_d + \varepsilon_N \right) + \frac{1}{4} K_e W_M^2}{\frac{1}{2} Q_m + K_r P_m + K_W W_M P_m}, \tag{8.5.13}$$

$$\|\tilde{W}(.)\|_F > \max \left\{ W_M, \frac{K_e W_M}{K_e - \frac{1}{32} \Lambda_M \gamma \sqrt{n}} \right\}.$$

Moreover, the filtered error $r(t)$ may be kept as small as desired by increasing the gain K_V in (8.5.8).

Proof. See the second appendix (section 8.B) of this chapter.

Corollary 8.5.2. *Let the conditions of Theorem 8.5.1 hold. Then $\delta(t)$, $e(t)$, $x(t)$, $\hat{x}(t)$, $\tau(t)$, and $u(t)$ are all UUB.*

Proof. Since from Theorem 8.5.1 $r(t)$ is UUB, then $\delta(t)$ is UUB by appropriately choosing Λ. Since $\delta(t) = x(t - T) + z(t) - z(t - 2T) - x_d(t)$ and $x_d(t)$ is bounded (Assumption 1), then $x(t - T) + z(t) - z(t - 2T)$ is UUB. Therefore, by definition, $u(t)$ is UUB for a stable A_S and a bounded $\dot{x}_d$. Using the boundedness of $\hat{W}(t)$ and $\tilde{x}(t)$, we can conclude that $x(t)$, $\hat{x}(t)$, $z(t)$, and $e(t)$ are also UUB. □

The way in which the RNN controller works is interesting. Note that from Theorem 8.5.1, the predictor approximates the local linearized subsystem with small error for a large K_r. This means the error between $z(t)$ and $x(t + T)$ or between $z(t - 2T)$ and $x(t - T)$ is small. Therefore, the predicted output of the plant $x(t)$ ahead of time by T seconds can be obtained via the predictor's output $z(t)$ with a small error. In this way, the effect of the delay can be eliminated and the tracking error can be kept small.

One of the main tools in control design for linear systems is the *separation principle*, which states that one may first design an observer for the state, and then, in a separate design, manufacture a control system assuming that the entire state is actually available as measurements. The separation principle states that when the controller is implemented using the estimated observed states, under certain conditions the overall closed-loop system is stable. Generally, a nonlinear separation principle is not valid. However, our design approach here along with the proofs shows that, for this telerobotic system, a separation principle indeed does hold. This allows the decoupling of the observer design and controller design stages.

Note that for the system identification stage in section 8.3, we required bounded inputs $\tau(t)$ and states $x(t)$ (Assumption 6). However, in the overall closed-loop design presented in this section, the requirement for bounded input and state have been replaced by bounds on the desired state $x_d(t)$ and $\dot{x}_d(t)$ (Assumption 1), which are reasonable in all practical systems.

The second term in the NN tuning law (8.5.10) corresponds to the e-mod (Narendra and Annaswamy (1987)), and the last term to the σ-mod (Ioannou and Kokotović (1984)). This allows us to avoid persistence of excitation conditions while guaranteeing bounded NN weights.

8.6 Simulation example

A two-link planar manipulator used extensively in the literature for illustration purposes appears in Figure 8.6.1. The dynamics of the two-link planar manipulator are given by Lewis, Abdallah, and Dawson (1993):

$$
\begin{bmatrix} a_1 + a_2 \cos q_2 & a_3 + \frac{a_2}{2} \cos q_2 \\ a_3 + \frac{a_2}{2} \cos q_2 & a_3 \end{bmatrix} \begin{bmatrix} \ddot{q}_1 \\ \ddot{q}_2 \end{bmatrix} + \begin{bmatrix} -a_2 \left(\dot{q}_1 \dot{q}_2 + \frac{\dot{q}_2^2}{2} \right) \sin q_2 \\ \frac{a_2}{2} \dot{q}_1^2 \sin q_2 \end{bmatrix}
$$
$$
+ \begin{bmatrix} a_4 \cos q_1 + a_5 \cos(q_1 + q_2) \\ a_5 \cos(q_1 + q_2) \end{bmatrix} + \begin{bmatrix} f_{v1} \dot{q}_1 + f_{c1} \operatorname{sgn}(\dot{q}_1) \\ f_{v2} \dot{q}_2 + f_{c2} \operatorname{sgn}(\dot{q}_2) \end{bmatrix} + \begin{bmatrix} \tau_{d1} \\ \tau_{d2} \end{bmatrix} = \begin{bmatrix} \tau_1 \\ \tau_2 \end{bmatrix}, \tag{8.6.1}
$$

where $a_1 = (m_1 + m_2)l_1^2 + m_2 l_2^2$, $a_2 = 2m_2 l_1 l_2$, $a_3 = m_2 l_2^2$, $a_4 = (m_1 + m_2)gl_1$, and $a_5 = m_2 g l_2$.

We took the arm parameters as $l_1 = 1$m, $l_2 = 0.8$m, $m_1 = 1$Kg, $m_2 = 1.2$Kg, $f_{v1} = f_{v2} = 2$, $f_{c1} = f_{c2} = 1.5$, and added an external disturbance of $\tau_{d1} = 5\sin(3t)$ and $\tau_{d2} = 5\cos(4t)$. The time delay in the input channel and the output channel was assumed to be $T = 0.5$ seconds. The proposed NN predictive control strategy was applied to this telerobot system for the trajectory tracking problem.

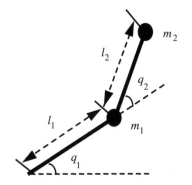

Figure 8.6.1. *Two-link planar elbow arm.*

Denote

$$y = q_1 + q_2 + \dot{q}_1 + \dot{q}_2. \tag{8.6.2}$$

The NN basis functions were chosen to be $\phi(x) = [\phi_1, \phi_2, \ldots, \phi_{19}]^T$ with components of

$$\phi_i(y) = \frac{e^{y+i-10} - e^{-y-i+10}}{e^{y+i-10} + e^{-y-i+10}}, \quad i = 1, 2, \ldots, 19, \tag{8.6.3}$$

where each $\phi_i(y)$ is a hyperbolic tangent function of y with bias along y.

The number of basis functions in the NN system heavily influences the complexity of NN systems. In general, the larger the number, the more complex the NN system and the higher the expected accuracy of the approximation. Hence there is always a tradeoff between complexity and accuracy in the choice of the number of basis functions. Their choice is usually quite subjective and based on some experience. In the above design, 19 hyperbolic tangent functions for variable y are chosen in which the bases for $i = 1, 2, \ldots, 19$ are selected as $-9, -8, \ldots, 8, 9$, respectively. Since the simulation below is in a neighborhood of the point $y = 0$, it is intuitively evident that the bases should be clustered around zero in the neighborhood (Figure 8.6.2).

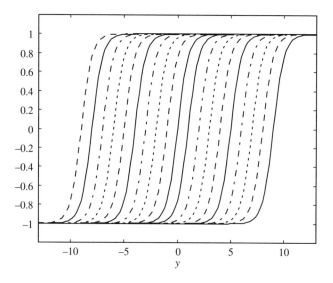

Figure 8.6.2. *NN basis functions $\phi_i(y)$.*

A simulation program was written in Visual C++ using a fourth-order fixed-stepsize Runge–Kutta integrator. The stepsize here was taken as 0.001 seconds. The initial conditions were chosen as $q_1(0) = 0$, $q_2(0) = 1$, $\hat{x}_1(0) = 0$, $\hat{x}_2(0) = 1$, $z_1(0) = 0$, $z_2(0) = 1$, and $\hat{W}(0) = 0$. We chose $A_{11} = \text{diag}\{-1, -1\}$ and $A_{12} = \text{diag}\{-2, -2\}$ so that A_S was stable. The control parameters were chosen as $F = \text{diag}\{50, \ldots, 50\}$, $\bar{\Lambda} = \text{diag}\{0.2, 0.2\}$, $K_e = 0.01$, $K_\sigma = 0.01$, $K_V = 20$, $K_r = 1$, and $K_W = 1$. The desired trajectory was selected as $q_{d1}(t) = \sin(t)$ and $q_{d2}(t) = \cos(t)$.

Remark. Larger K_r may yield better performance in attenuating the effect of the estimate error and larger K_V may yield better tracking performance, but the control signal during the transient time is larger than the case of smaller K_V. Hence there is a tradeoff between the tracking performance and the amplitude of the control signal during the transient time.

Simulation results are shown in Figures 8.6.3–8.6.8. The tracking performances are presented in Figures 8.6.3 and 8.6.4, where the desired and actual trajectories are shown. Also shown is the delayed desired trajectory. Note that, due to the communication channel delay, the actual trajectory follows the delayed desired trajectory. The approximation errors between the delayed desired trajectory and the actual trajectory are presented in Figure 8.6.5. The filtered errors $r(t)$ are plotted in Figure 8.6.6, and the control signals $u(t)$ and $\tau(t)$ are depicted in Figures 8.6.7 and 8.6.8, respectively.

We can compare the performance of $q_1(t)$ and $q_2(t)$ with the delayed reference states $q_{d1}(t-T)$ and $q_{d2}(t-T)$, respectively, because of the time delay in the input channel. From

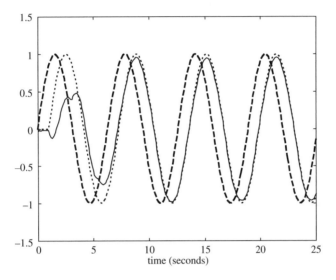

Figure 8.6.3. *Response of the angular position $q_1(t)$: Solid, $q_1(t)$; dashed, $q_{d1}(t)$; dotted, $q_{d1}(t-0.5)$.*

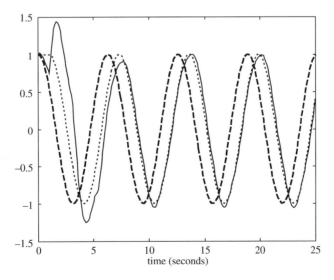

Figure 8.6.4. *Response of the angular position $q_1(t)$: Solid, $q_2(t)$; dashed, $q_{d2}(t)$; dotted, $q_{d2}(t-0.5)$.*

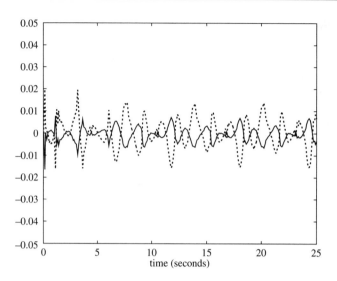

Figure 8.6.5. *Responses of the RNN approximation errors: Solid, $\tilde{q}_1$; dotted, $\tilde{q}_2$.*

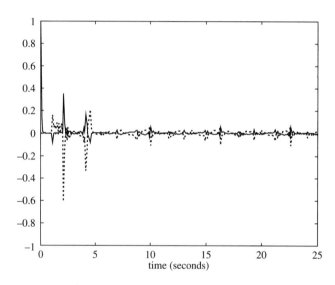

Figure 8.6.6. *Response of the filtered error $r(t)$ Solid, $r_1(t)$; dotted, $r_2(t)$.*

Figures 8.6.3 and 8.6.4, we can see the proposed control strategy can cancel the delay effect in the closed-loop system and has good tracking performance despite the communication channel delay.

This control system makes the telerobot follow the delayed desired trajectory no matter how large the delay T.

8.7   Conclusion

A new NN control structure has been proposed for the control of telerobotic system with constant time delays caused by communication channels. The NN controller extends the

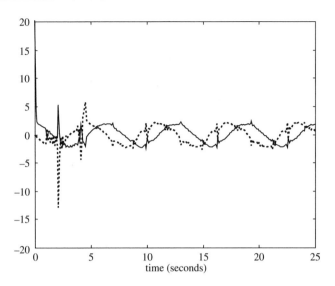

Figure 8.6.7. *Response of the remote input $u(t)$: Solid, $u_1(t)$; dotted, $u_2(t)$.*

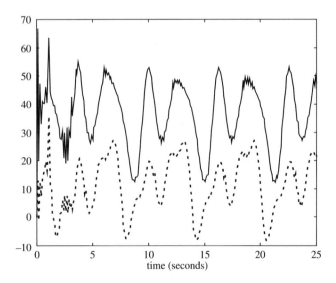

Figure 8.6.8. *Response of the torque symbol $\tau(t)$: Solid, $\tau_1(t)$; dotted, $\tau_2(t)$.*

Smith predictor idea to nonlinear robotic systems. A rigorous stability analysis has been performed based on the Lyapunov method. Satisfactory tracking performance of the telerobot control system has been verified through numerical simulations.

8.A   Appendix: Proof of Theorem 8.3.1

Select the Lyapunov function candidate

$$V = \frac{1}{2}\tilde{x}^T P \tilde{x} + \frac{1}{2}\operatorname{tr}(\tilde{W}^T F^{-1} \tilde{W}). \tag{8.A.1}$$

Differentiating yields

$$\dot{V} = \frac{1}{2}(\dot{\tilde{x}}^T P \tilde{x} + \tilde{x}^T P \dot{\tilde{x}}) + \text{tr}\{\tilde{W}^T F^{-1} \dot{\tilde{W}}\}, \tag{8.A.2}$$

whence substituting now from the error system (8.3.5) and Lyapunov equation (8.2.6) yields

$$\dot{V} = -\frac{1}{2}\tilde{x}^T Q \tilde{x} - \tilde{x}^T P B M^{-1}(x)\tau_d + \tilde{x}^T P \varepsilon - \tilde{x}^T P v + \text{tr}\{\tilde{W}^T (F^{-1}\dot{\tilde{W}} + \phi(x)\tilde{x}^T P b)\}. \tag{8.A.3}$$

The tuning law (8.3.6) and robust term (8.3.7) give (recall that $\dot{\tilde{W}} = -\dot{\hat{W}}$)

$$\dot{V} = -\frac{1}{2}\tilde{x}^T Q \tilde{x} + \tilde{x}^T P B M^{-1}(x)\tau_d + \tilde{x}^T P \varepsilon - K_r \tilde{x}^T P \tilde{x} - K_W \|\hat{W}\|_F \tilde{x}^T P \tilde{x} \\ + K_e \|\tilde{x}\| \text{tr}\{\tilde{W}^T \hat{W}\}. \tag{8.A.4}$$

Since

$$\text{tr}\{\tilde{W}^T \hat{W}\} = \text{tr}\{\tilde{W}^T (W - \tilde{W})\} = \langle \tilde{W}, W \rangle - \|\tilde{W}\|_F^2 \tag{8.A.5}$$

and P is a positive definite matrix, there results

$$\dot{V} \le -\frac{1}{2}Q_m \|\tilde{x}\|^2 + P_M\left(\sqrt{n}m_1^{-1}b_d + \varepsilon_N\right)\|\tilde{x}\| + K_e\|\tilde{x}\|(W_M\|\tilde{W}\|_F - \|\tilde{W}\|_F^2) - K_r P_m\|\tilde{x}\|^2$$

$$\le -\|\tilde{x}\|\left\{ \left(\frac{1}{2}Q_m + K_r P_m\right)\|\tilde{x}\| + K_e\left(\|\tilde{W}\|_F - \frac{1}{2}W_M\right)^2 \right. \tag{8.A.6}$$

$$\left. - \left(P_M\left(\sqrt{n}m_1^{-1}b_d + \varepsilon_N\right) + \frac{1}{4}K_e W_M^2\right)\right\}$$

with $Q_m = \lambda_{\min}(Q)$ and $P_m = \lambda_{\min}(P)$ the smallest eigenvalues of Q and P, respectively, and $P_M = \lambda_{\max}(P)$ the largest eigenvalue of P. Thus $\dot{V}$ is negative as long as either

$$\|\tilde{x}\| > \frac{P_M\left(\sqrt{n}m_1^{-1}b_d + \varepsilon_N\right) + \frac{1}{4}K_e W_M^2}{\frac{1}{2}Q_m + K_r P_m} \tag{8.A.7}$$

or

$$\|\tilde{W}\|_F > \frac{1}{2}W_M + \sqrt{\frac{P_M\left(\sqrt{n}m_1^{-1}b_d + \varepsilon_N\right)}{K_e} + \frac{1}{4}W_M^2}. \tag{8.A.8}$$

Therefore, $\dot{V}$ is negative outside a compact set in the $\|\tilde{x}\|$, $\|\tilde{W}\|_F$ plane, which is thus shown to be an attractive set for the system. According to a standard Lyapunov theorem extension, this demonstrates the UUB of both $\tilde{x}$ and $\hat{W}$. $\square$

8.B   Appendix: Proof of Theorem 8.5.1

Consider the Lyapunov function candidate

$$V = \frac{1}{2}r^T(t)r(t) + \frac{1}{2}\tilde{x}^T(t-T)P\tilde{x}(t-T) + \frac{1}{2}\text{tr}\{\tilde{W}^T(t-T)F^{-1}\tilde{W}(t-T)\}$$

$$+ \frac{1}{2}\tilde{x}^T(t)P\tilde{x}(t) + \frac{1}{2}\{\tilde{W}^T(t)F^{-1}\tilde{W}(t)\}. \tag{8.B.1}$$

Differentiating yields

$$\dot{V} = r(t)\dot{r}(t) + \frac{1}{2}\{\dot{\tilde{x}}^T(t-T)P\tilde{x}(t-T) + \tilde{x}^T(t-T)P\dot{\tilde{x}}(t-T)\}$$

$$+ \text{tr}\{\tilde{W}^T(t-T)F^{-1}\dot{\tilde{W}}(t-T)\} + \frac{1}{2}\{\dot{\tilde{x}}^T(t)P\tilde{x}(t) + \tilde{x}^T(t)P\dot{\tilde{x}}(t)\} \qquad (8.B.2)$$

$$+ \text{tr}\{\tilde{W}^T(t)F^{-1}\dot{\tilde{W}}(t)\}.$$

Equation (8.B.2) is evaluated along error equations (8.3.5) and (8.5.9). Then applying Lyapunov equations (8.2.6) gives

$$\dot{V} = -r^T(t)K_V r(t)$$

$$- r(t)\Lambda^T\{B\tilde{W}^T(t-T)\phi(x(t-T)) + BM^{-1}(x(t-T))\tau_d(t-T) + \varepsilon(t-T)\}$$

$$- \frac{1}{2}\tilde{x}^T(t-T)Q\tilde{x}(t-T)$$

$$+ \tilde{x}^T(t-T)\{-PBM^{-1}(x(t-T))\tau_d(t-T) + P\varepsilon(t-T)\}$$

$$- \tilde{x}^T(t-T)Pv(t-T) \qquad (8.B.3)$$

$$+ \text{tr}\{\tilde{W}^T(t-T)(F^{-1}\dot{\tilde{W}}(t-T) + \phi(x(t-T))\tilde{x}^T(t-T)PB)\}$$

$$- \frac{1}{2}\tilde{x}^T(t)Q\tilde{x}(t) + \tilde{x}^T(t)\{-PBM^{-1}(x(t))\tau_d(t) + P\varepsilon(t)\} - \tilde{x}^T(t)Pv(t)$$

$$+ \text{tr}\{\tilde{W}^T(t)(F^{-1}\dot{\tilde{W}}(t) + \phi(x(t))\tilde{x}^T(t)PB)\}.$$

Using (8.5.10) and (8.5.11) yields

$$\dot{V} = -r^T(t)K_V r(t)$$

$$- r(t)\Lambda^T\{B\tilde{W}^T(t-T)\phi(x(t-T)) + BM^{-1}(x(t-T))\tau_d(t-T) + \varepsilon(t-T)\}$$

$$- \frac{1}{2}\tilde{x}^T(t-T)Q\tilde{x}(t-T)$$

$$+ \tilde{x}^T(t-T)\{-PBM^{-1}(x(t-T))\tau_d(t-T) + P\varepsilon(t-T)\}$$

$$- K_r\tilde{x}^T(t-T)P\tilde{x}(t-T) - K_W\|\hat{W}(t-T)\|_F\tilde{x}^T(t-T)P\tilde{x}(t-T) \qquad (8.B.4)$$

$$+ K_e\|\tilde{x}(t-T)\|\,\text{tr}\{\tilde{W}^T(t-T)\hat{W}(t-T)\} + K_\sigma\,\text{tr}\{\tilde{W}^T(t-T)\hat{W}(t-T)\}$$

$$- \frac{1}{2}\tilde{x}^T(t)Q\tilde{x}(t) + \tilde{x}^T(t)\{-PBM^{-1}(x(t))\tau_d(t) + P\varepsilon(t)\} - K_r\tilde{x}^T(t)P\tilde{x}(t)$$

$$- K_W\|\hat{W}(t)\|_F\tilde{x}^T(t)P\tilde{x}(t) + K_e\|\tilde{x}(t)\|\,\text{tr}\{\tilde{W}^T(t)\hat{W}(t)\} + K_\sigma\,\text{tr}\{\tilde{W}^T(t)\hat{W}(t)\}.$$

Using inequality (8.A.5), we have

$$\dot{V} \leq -K_V\|r(t)\|^2 + \Lambda_M\gamma\sqrt{n}\|r(t)\|\|\tilde{W}(t-T)\|_F + \Lambda_M\|r(t)\|\left(\sqrt{n}m_1^{-1}b_d + \varepsilon_N\right)$$

$$- \frac{1}{2}Q_m\|\tilde{x}(t-T)\|^2 + \|\tilde{x}(t-T)\|P_M\left(\sqrt{n}m_1^{-1}b_d + \varepsilon_N\right) - K_r P_m\|\tilde{x}(t-T)\|^2$$

$$- K_W W_M P_m\|\tilde{x}(t-T)\|^2 - K_W P_m\|\tilde{W}(t-T)\|_F\|\tilde{x}(t-T)\|^2$$

$$+ K_e\|\tilde{x}(t-T)\|\{W_M\|\tilde{W}(t-T)\|_F - \|\tilde{W}(t-T)\|_F^2\}$$

$$+ K_\sigma\{W_M\|\tilde{W}(t-T)\|_F - \|\tilde{W}(t-T)\|_F^2\} - \frac{1}{2}Q_m\|\tilde{x}(t)\|^2 \qquad (8.B.5)$$

$$+ \|\tilde{x}(t)\|P_M\left(\sqrt{n}m_1^{-1}b_d + \varepsilon_N\right) - K_r P_m\|\tilde{x}(t)\|^2 - K_W W_M P_m\|\tilde{x}(t)\|^2$$

$$- K_W P_m\|\tilde{W}(t)\|_F\|\tilde{x}(t)\|^2 + K_e\|\tilde{x}(t)\|\{W_M\|\tilde{W}(t)\|_F - \|\tilde{W}(t)\|_F^2\}$$

$$+ K_\sigma\{W_M\|\tilde{W}(t)\|_F - \|\tilde{W}(t)\|_F^2\},$$

where $\Lambda_M = \lambda_{\max}(\bar{\Lambda})$ is the largest eigenvalue of $\bar{\Lambda}$ and the inequality $\|\hat{W}\|_F \leq W_M + \|\tilde{W}\|_F$ has been used.

It is easy to verify that for real numbers k, a, and b, the inequality

$$ab \leq \frac{a^2 + k^2 b^2}{2k} \tag{8.B.6}$$

is true. Thus after manipulating (8.B.5), we get

$$
\begin{aligned}
\dot{V} \leq &-\|r(t)\| \left\{ \left(K_V + 8\Lambda_M \gamma \sqrt{n}\right) \|r(t)\| - \Lambda_M \left(\sqrt{n} m_1^{-1} b_d + \varepsilon_N\right) \right\} \\
&- \|\tilde{x}(t-T)\| \left\{ \left(\frac{1}{2} Q_m + K_r P_m + K_W W_M P_m\right) \|\tilde{x}(t-T)\| \right. \\
&\qquad\qquad \left. - K_e W_M \|\tilde{W}(t-T)\|_F + K_e \|\tilde{W}(t-T)\|_F^2 - P_M \left(\sqrt{n} m_1^{-1} b_d + \varepsilon_N\right) \right\} \\
&- \|\tilde{x}(t)\| \left\{ \left(\frac{1}{2} Q_m + K_r P_m + K_W W_M P_m\right) \|\tilde{x}(t)\| \right. \\
&\qquad\qquad \left. - K_e W_M \|\tilde{W}(t)\|_F + K_e \|\tilde{W}(t)\|_F^2 - P_M \left(\sqrt{n} m_1^{-1} b_d + \varepsilon_N\right) \right\} \\
&- \|\tilde{W}(t-T)\|_F \left\{ \left(K_\sigma - \frac{1}{32} \Lambda_M \gamma \sqrt{n}\right) \|\tilde{W}(t-T)\|_F - K_\sigma W_M \right\} \\
&- K_\sigma \|\tilde{W}(t)\|_F \{\|\tilde{W}(t)\|_F - W_M\} - K_W P_m \|\tilde{W}(t-T)\|_F \|\tilde{x}(t-T)\|^2 \\
&- K_W P_m \|\tilde{W}(t)\|_F \|\tilde{x}(t)\|^2.
\end{aligned}
\tag{8.B.7}
$$

Completing the square terms in (8.B.7) yields that the right side of (8.B.7) is negative if, in addition to (8.5.12), condition (8.B.8) holds:

$$
\begin{aligned}
\|r(t)\| &> \frac{\Lambda_M \left(\sqrt{n} m_1^{-1} b_d + \varepsilon_N\right)}{K_V + 8\Lambda_M \gamma \sqrt{n}}, \\
\|\tilde{x}(.)\| &> \frac{P_M \left(\sqrt{n} m_1^{-1} b_d + \varepsilon_N\right) + \frac{1}{4} K_e W_M^2}{\frac{1}{2} Q_m + K_r P_m + K_W W_M P_m}, \\
\|\tilde{W}(.)\|_F &> \max \left\{ W_M, \frac{K_e W_M}{K_e - \frac{1}{22} \Lambda_M \gamma \sqrt{n}} \right\}.
\end{aligned}
\tag{8.B.8}
$$

From (8.B.1), we can see that V is a positive definite function. Since $\dot{V}$ is negative definite for all $r(t)$, $\tilde{x}(t)$, and $\tilde{W}(t)$ satisfying (8.5.12) and (8.B.8), we conclude that $r(t)$, $\tilde{x}(t)$, and $\tilde{W}(t)$ are UUB. $\square$

Chapter 9

Implementation of Neural Network Control Systems

Guest authors: S. Ikenaga and J. Campos

This chapter shows how to implement NN controllers on actual industrial systems. A real-time control system developed at the Automation and Robotics Research Institute (ARRI) of the University of Texas at Arlington (UTA) is described. The hardware and the software interfaces are outlined. Next, a real-world problem is described—namely, control of a military tank gun barrel. This vibratory flexible system presents severe difficulties for high-speed precision pointing control. A dynamical mathematical model of a flexible-link pointing system is developed, and an NN controller is derived for accurate pointing control with vibration rejection. Finally, it is described how the NN controller was implemented on the ATB1000 tank gun barrel testbed at the U.S. Army Armament Research, Development, and Engineering Center, Picatinny, NJ.

9.1   PC–PC real-time digital control system

Implementation of advanced or otherwise nonstandard control algorithms has long been an expensive and time-consuming task. Many commercial products are available that implement traditional PID servocontrol, but anything other than PID control often requires the use of custom digital signal processor boards and development systems. The development costs can be prohibitive for projects with modest budgets. A variety of systems that traditionally use PID control can potentially benefit from advanced control algorithms. Among these are inertial stabilization of gun platforms and antenna pointing systems, and control of flexible mechanical structures.

The real-time control system (RTCS) was developed at UTA's ARRI to facilitate the implementation of advanced control algorithms. RTCS was developed under Small Business Innovation Research (SBIR), National Science Foundation (NSF), and Texas Advanced Technology Program (ATP) funding. Based on two inexpensive PC computers and nicknamed "PC–PC," its premise is simple: use one computer (target computer) for real-time control, and use a second computer (host computer) for the user interface to the controller. Both computers are industry standard IBM® PC compatibles. The computers can be standard desktop models or industrially rugged. The cost of the computers is relatively cheap, a major consideration in the design of RTCS.

The division of labor between two computers (see Figure 9.1.1) allows the use of an operating system that is most suitable to the task at hand for each subsystem. The VRTXsa86™ real-time operating system (RTOS) from Mentor Graphics® is used on the target computer, while Microsoft® Windows® 2000 Professional is used on the host computer. The computers communicate via the Transmission Control Protocol/Internet Protocol (TCP/IP) either through a single cable directly connected between the two computers, or through a local-area network (LAN) or wide-area network (WAN). Since TCP/IP is the Internet transport protocol, the Internet itself or a wireless LAN can be used as a communication link.

The data acquisition on the target computer is accomplished using IndustryPack® (IPack) products from SBS® GreenSpring Modular I/O and Systran® Corporation. IPack modules support simple input/output (I/O) for analog and digital data. Each IPack is about the size of a business card. Four to six IPacks plug onto a carrier board that contains the host bus interface and external cable connections. Carrier boards have been implemented for several different types of bus systems such as 3U VME, 6U VME, NuBus™, ISA, VXI, PCI, and Compact PCI®.

The PC–PC controller simplifies implementation of advanced control algorithms and allows an unprecedented level of real-time system monitoring and interaction with the user. Any signal processed by the controller is accessible to the host computer at any time. All control system gains and configuration parameters can be viewed and modified while the controller is operating. Development of ARRI's PC–PC controller provided the means necessary to test the active suspension control design discussed in Chapter 3 on DTI's quarter-vehicle testbed.

This chapter presents an overview of the PC–PC RTCS along with a description of the hardware and software architectures. More information is available in Ikenaga (2000).

9.1.1   Hardware description

Figure 9.1.1 depicts the overall PC–PC RTCS architecture. The target PC consists of an industry standard Pentium® CPU-based ISA bus computer with 16 MBs of random-access memory (RAM). A standard 3.5-inch high-density floppy disk is all that is required to boot the complete real-time application onto the target PC. A network interface card (NIC) is also required to communicate with the host PC via TCP/IP. The RTOS vendor recommends an NIC compatible with either a Novell NE2000™ or 3COM® Etherlink®-III. Although not required, a standard VGA video graphics card is also included.

The data acquisition (DAQ) boards provide the I/O signals from the plant. Such signals consist of analog or digital I/O from sensors (e.g., accelerometers, potentiometers, strain gauges, and encoders) and actuators (e.g., motors, relays, and amplifiers). The PC–PC RTCS uses the IndustryPack technology which has become a de facto standard due to the successful MVME162 board from Motorola®. The two basic components of the DAQ boards are the IP-module and its carrier board.

An example of an IP-module is shown in Figure 9.1.2. Approximately the size of a business card, each IP-module provides a high I/O density in a modular package. IP-modules exist in two sizes: single and double size, where the double size IP-module is seen as two single-size IP-modules together. The single size IP-module uses two 50-pin connectors: one with the plant (via I/O interface wiring blocks) and one to interface with the carrier board. The RTCS employs Systran's ADC128F1 IP-module for eight channels of 12-bit analog input-signals simultaneously, Systran's DAC128V IP-module for eight channels of 12-bit analog-outputs, and GreenSpring's IP-Quadrature IP-module for four channels of encoder inputs. All three IP-modules reside on a carrier board like the one shown in Figure 9.1.3.

Systran's ISASC6 board shown in Figure 9.1.3 is a six-slot IP carrier board that supports six single-size or three double-size IP-modules. It uses a standard ISA (PC/AT) 8/16-bit bus interface and provides up to 300 I/O points (50 per IP-module) per ISA slot within the target PC. Since the IP-module is passive on the bus, it can only be accessed via the carrier board.

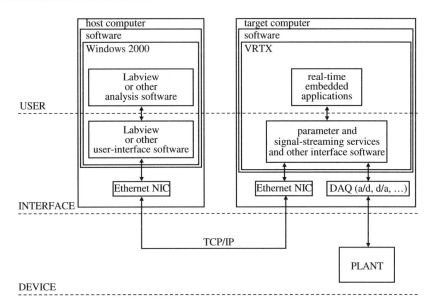

Figure 9.1.1. *PC–PC RTCS architecture.*

Figure 9.1.2. *IndustryPack module.* (*Used with permission of SBS Technologies, Inc.*)

Figure 9.1.3. *Systran's ISASC6 six-slot IP carrier board.* (*Used with permission of SBS Technologies, Inc.*)

A point-to-point IP-bus from the carrier board to IP-module provides for the multiplexed address and data bus. Thus each carrier board provides a bridge between the ISA bus and IP-bus. Fundamentally IndustryPack products provide flexibility by allowing one to choose specific types of IP-modules necessary for a specific plant and the type of bus architecture (i.e., ISA, PCI, and CPCI).

The host PC will typically be configured with the latest PC technology available in the market running an operating system such as Windows 2000 Professional. Additional software like National Instruments™ LabVIEW or Microsoft Visual Basic® is required to provide for the host-user application software. The host-user applications provide the real-time system monitoring and interaction with the user. Any signal processed by the controller can be viewed in real time while controller gains and configuration parameters are modified online.

9.1.2   Software description

This section describes the RTOS used on the PC–PC RTCS. In addition, several fundamental software architectures such as the networking communication services, I/O data management, and state-variable data management are given.

The VRTX® real-time operating system

The VRTX (versatile real-time executive) RTOS is a kernel for embedded system projects requiring hard real-time responsiveness, multitasking capability, and system modularity. The VRTX x 86/fpm Developer's Kit consisting of the VRTX RTOS and a complete set of compatible PC-hosted development tools (e.g., C/C++ compilers, debuggers, macro assembler, linker/locator, and librarian) facilitates development of embedded applications on all 32-bit x 86 CPUs (e.g., 80 x 386, 80 x 486, and Pentium processors). Its modularity enables VRTX to scale to the needs of the user's application and memory requirements. VRTX is certifiable under the United States' Federal Aviation Administration (FAA) RTCA/DO-178B level A standards for mission-critical aerospace systems. Complete details on VRTX x 86/fpm can be found in Lee and Kim (1994).

The PC–PC RTCS networking is done through the STREAMS Networking Executive (SNX™) module provided as an additional option by Mentor Graphics. SNX provides a System V STREAMS framework and a STREAMS-based TCP/IP application and protocol set for VRTX. Device drivers for NICs compatible with the Novell NE2000 or 3COM Etherlink-III have also been provided in conjunction with SNX to allow easier development of higher level network application programming. The STREAMS-based protocol stack facilitates the communication between the host and target PCs that allows the user to view and modify control system gains and configuration parameters while viewing real-time signals of sensors and actuators.

Networking communication services

The PC–PC RTCS networking communication architecture is shown in Figure 9.1.4. The PC–PC Super Server provides the functions necessary to establish and manage host-target connections over TCP/IP. It initializes and configures all PC–PC communication services by reading parameters from a user-supplied initialization string, starts the PC–PC Memory Manager and Service Manager, and installs the user-specified PC–PC services. After completing all initialization and configuration tasks, the PC–PC Super Server waits for client-connections from a host PC.

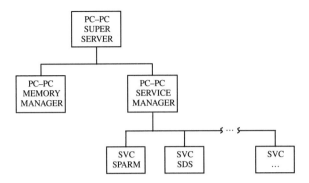

Figure 9.1.4. *PC–PC networking architecture.*

The PC–PC Memory Manager creates and manages the memory heap used by the PC–PC communication services. This module must be successfully executed before any other initialization or configuration task can be completed because memory for the list of installed services is allocated from this memory heap. Additionally, services such as the Simple Parameter Access Service (SVC_SPARM) and the Simple Data Streaming Service (SVC_SDS) utilize circular buffers allocated from this memory heap in order to pass control parameters and signals between the host and target PCs. The PC–PC Memory Manager properly allocates and manages these buffers.

The PC–PC Service Manager manages the PC–PC communication services by maintaining a list of user-specified installed services. It allocates memory for each communication service by calling the PC–PC Memory Manager, sets the maximum number of clients per service and maximum number of services allowable by the PC–PC Super Server, and starts a communication service when a client opens a connection. The type of communication service depends upon the service identification (SVCID) sent from the client (host PC).

The types of PC–PC communication services currently available are the SVC_SPARM, SVC_SDS, System Log Streaming Service (SVC_LOG), and CPU Usage Monitoring Service (SVC_LOG). The SVC_SPARM service allows the user to modify controller gains and configuration parameters. It should be noted that for fail-safe reasons, this service is the only one that permits only a single user to modify parameters at any given moment. Thus other clients must wait until the current client closes the SVC_SPARM. The SVC_SDS service streams all signals processed by the controller to be viewed by the user in real time. The SVC_LOG service provides a connection log of the clients connected to the target PC. The SVC_CPU service is very useful, especially during the debugging stage, in determining the amount of CPU processing power used by the current application.

Input/output data management

The input and output of actual signals from the plant (e.g., sensors and actuators) is organized as shown in Figure 9.1.5. Each DAQ board device that exists within the target PC requires its own device driver. Device drivers are low-level software modules that provide the means for high-level RTOS applications (user) to access physical devices in the PC (e.g., parallel ports, keyboards, mouse, and DAQ boards). Typical DAQ board device drivers provide critical functions for board configuration, initialization, and signal input and output through an application programming interface (API) that determines how to properly access the board's data and address registers to accomplish each function. Thus a user accesses plant

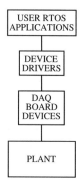

Figure 9.1.5. *I/O data management architecture.*

signals from a specific DAQ board from function calls defined in the API.

9.1.3   State-variable data management

This section describes the management of the state-variable structure of the PC–PC RTCS. In order to implement a control algorithm on the RTCS, a parameter data structure and signal data structure are required to manage the parameter configurations and data signals being accessed by a client through the networking communication services. A detailed explanation of PC–PC application development can be found in Scully (1998).

The control block diagram

The basis of each real-time digital controller implementation on the PC–PC RTCS is the control block diagram like the one shown in Figure 9.1.6. Similar in form and function to a control diagram found in the control literature, the PC–PC diagram also includes additional blocks that pertain to DAQ board devices that interface with I/O signals from the plant. Figure 9.1.6 illustrates how actual I/O signals from the plant become state variables within the RTOS application. This is shown in the figure as two regions: virtual and physical, divided by the dashed line.

There are two types of blocks shown in Figure 9.1.6: an I/O block (IOB) and a signal block (SB). The IOB, shown at the interface between the physical and virtual regions,

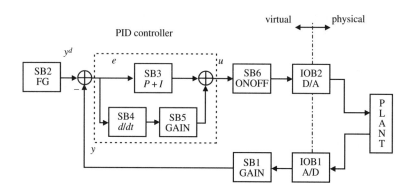

Figure 9.1.6. *PC–PC control block diagram.*

establishes where input and output signals from the plant are passed to the target PC via DAQ board device drivers. As mentioned previously, device drivers are low-level software modules that provide the means for higher level applications to access physical devices in the PC. The SB is a virtual signal-processing block that generates a specified output given by the block's name. The name of the SB describes the function that the SB performs. Table 9.1.1 lists the various names of SBs available to the user. The SB facilitates implementation of the control algorithm in the RTOS application.

Table 9.1.1. *SB names for PC–PC RTCS.*

Block descriptor	Block description
sb_but1lp	First-order Butterworth low-pass filter (LPF)
sb_but2bp	Second-order Butterworth band-pass filter (BPF)
sb_but2lp	Second-order Butterworth LPF
sb_but2n	Second-order Butterworth notch filter (NF)
sb_but3lp	Third-order Butterworth LPF
sb_but4lp	Fourth-order Butterworth LPF
sb_but4n	Fourth-order Butterworth NF
sb_deriv	Differentiator
sb_fg	Function generator
sb_friccomp	Friction compensation
sb_gain	Proportional gain
sb_ident	Identity filter
sb_integ	Integrator
sb_integhp	High-pass integral filter for accelerometers
sb_nulladj	Null adjust
sb_onoff	On–off
sb_pi	Proportional plus integral gain with antiwindup
sb_sat	Saturation
sb_tg2	Trajectory generator (finite second-order derivative)
sb_trint	Trapezoidal integrator
sb_volt2str	Volt-to-strain conversion

Parameter data structure

The client user has access to control parameters through the SVC_SPARM networking communication service. This service utilizes a fixed-size buffer to store and transmit control parameters to the client and read parameter modifications from the client. The parameters are organized in a parameter data structure that contains an ordered list of each SB and their parameters. To prevent the user from making any unnecessary errors while running the controller, all IOBs are omitted from this data structure. The IOBs are properly configured before the application code is compiled. An example of the parameter data structure for the control block diagram in Figure 9.1.6 is shown in Figure 9.1.7. As shown, every SB contains a "mode" parameter that enables the user to turn the block on or off and contains

SB1	SB2	SB3	SB4	SB5	SB6
mode	mode	mode	mode	mode	mode
gain	amp	kp	freq	gain	enable
	freq	ki			
	wave				

Figure 9.1.7. *Parameter data structure.*

additional parameters specific to the functionality of the SB. Additional modes of operation and parameters associated with each SB are discussed in Scully (1998).

Signal data structure

The SVC_SDS service provides real-time streaming of all signals from the target PC to the host PC. The signal data structure scheme is shown in Figure 9.1.8. A circular buffer, known as the signal data buffer, stores the signals at each sample period. The buffer size is fixed and predetermined by the user. If there are no more empty slots to store a new signal data structure, then the service writes the new data into the oldest time slot in the buffer. Each time slot contains a signal data structure consisting of each signal from IOBs and SBs as shown in Figure 9.1.8.

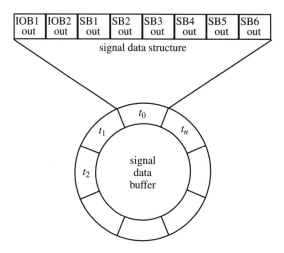

Figure 9.1.8. *Signal data structure scheme.*

9.2   ATB1000 U.S. Army tank gun barrel testbed

We intend to show how to use RTCS to implement an NN controller on a U.S. Army tank gun barrel testbed. A real tank like the U.S. Army M1A1 (Figure 9.2.1), one of the key land vehicles in the Gulf War in 1991, is a 64-ton behemoth protected by a steel-encased depleted uranium armor. It has a rigid 120 mm M256 barrel and a powerful 1500 hp engine for moving

Figure 9.2.1. *U.S. Army M 1A1 tank.*

and aiming purposes. Besides the fact that the barrel has to withstand the firing, it is made rigid to avoid flexibility when pointing at high speed. However, the engine is so powerful that it could eventually turn the tank completely sideways if the velocity during pointing is not properly controlled. This represents a drawback when facing multiple targets at a time. Unmodeled system dynamics such as friction prevent one from using a fully automatic control system in the aiming process. Human intervention is necessary to correct for errors.

In future models of military tanks, a lighter material could potentially be used for the gun barrel. This would reduce the problems associated with a massive barrel, but it would require a far more competent automatic control system. An intelligent control system could compensate for barrel tip flexibility and unmodeled dynamics. Such an intelligent control system would allow control of flexible lightweight gun barrels, and one such controller, based on NNs, is proposed in this chapter. This would allow higher barrel traverse pointing speeds and better accuracy when aiming and would significantly reduce the need for human intervention.

The ATB1000 testbed at the U.S. Army's ARDEC in Picatinny, NJ models a gun barrel/turret system with a flexible barrel. Under an SBIR contract, an NN pointing controller was implemented on the ATB1000. In this chapter, we describe how it was done using RTCS.

The ATB1000 has highly complex nonlinearities such as friction, backlash, and hysteresis. These nonlinearities in practice are found in almost every mechanical actuator. A schematic of the ATB1000 is shown in Figure 9.2.2. Note it has a compliant drive train coupling and a flexible gun barrel. The control problem is to control the position of the tip of the barrel, but only the angles (azimuth and elevation) at the hub may be directly controlled. The tip position is deformed by the vibrational modes.

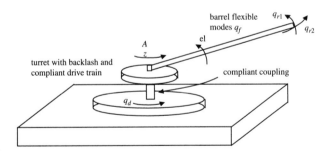

Figure 9.2.2. *U.S. Army ATB1000 tank gun barrel testbed.*

9.3 Derivation of neural net control system for flexible systems

The control of mechanical systems with flexibility and vibratory modes is not an easy task. One must achieve the prescribed motion and pointing behavior of the desired controlled variables while simultaneously suppressing unwanted vibratory modes in all motion variables. This is a problem in two time scales: the fast vibratory motion and the slower desired motion. Here we show how to use the two-time-scale singular perturbation approach (Kokotović (1984)) to design an NN controller to accurately point the ATB1000 tank gun barrel while suppressing vibratory modes.

The NN controller derived in this section appears in Figure 9.3.1. More details on the

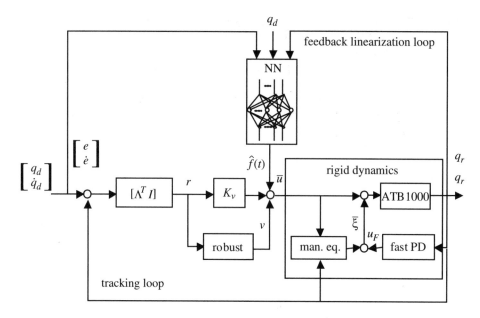

Figure 9.3.1. *Control structure of the NN controller for a flexible-link pointing system.*

NN controller derivation are given in Lewis, Jagannathan, and Yesildirek (1999).

9.3.1   Flexible-link robot dynamics

The dynamics of a multilink flexible-link robot can be represented as

$$M(q)\ddot{q} + D(q,\dot{q})\dot{q} + Kq + F(q,\dot{q}) + G(q) = B(q)u, \qquad (9.3.1)$$

where $q = [q_r \quad q_f]^T$, q_r is the vector of rigid modes, and q_f is the vector of flexible modes. $M(q)$ is the inertia matrix, $D(q,\dot{q})$ is the Coriolis and centripetal matrix, K is the stiffness matrix, $F(q,\dot{q})$ is the friction matrix, $G(q)$ is the gravity matrix, $B(q)$ is an input matrix dependent on the boundary conditions selected in the assumed mode-shaped method, and u includes the control torques applied to each joint. The dynamics (9.3.1) can be partitioned as

$$
\begin{aligned}
M_{rr}\ddot{q}_r + M_{rf}\ddot{q}_f + V_{rr}\dot{q}_r + V_{rf}\dot{q}_f + F_r(\dot{q}_r) + G_r(q_r) &= B_r u_r, \\
M_{fr}\ddot{q}_f + M_{ff}\ddot{q}_f + V_{fr}\dot{q}_r + V_{ff}\dot{q}_f + K_{ff}q_f &= B_f u_f.
\end{aligned}
\qquad (9.3.2)
$$

Equation (9.3.2) can be further expressed as

$$
\begin{aligned}
\ddot{q}_r &= -V_{rr}^1 \dot{q}_r - V_{rf}^1 \dot{q}_f - K_{rf}^1 q_f - F_r^1 - G_r^1 + B_r^1 u_r, \\
\ddot{q}_f &= -V_{fr}^1 \dot{q}_r - V_{ff}^1 \dot{q}_f - K_{ff}^1 q_f - F_f^1 - G_f^1 + B_f^1 u_r,
\end{aligned}
\qquad (9.3.3)
$$

where

$$\begin{bmatrix} H_{rr} & H_{rf} \\ H_{fr} & H_{ff} \end{bmatrix} = \begin{bmatrix} M_{rr} & M_{rf} \\ M_{fr} & M_{ff} \end{bmatrix}^{-1},$$

$$V_{rr}^1 = H_{rr} V_{rr} + H_{rf} V_{fr}, \qquad V_{fr}^1 = H_{fr} V_{rr} + H_{ff} V_{fr},$$

$$V_{rf}^1 = H_{rr} V_{rf} + H_{rf} V_{ff}, \qquad V_{ff}^1 = H_{fr} V_{rf} + H_{ff} V_{ff},$$

$$K_{rf}^1 = H_{rf} K_{ff}, \qquad K_{ff}^1 = H_{ff} K_{ff}, \tag{9.3.4}$$

$$F_r^1 = H_{rr} F_r, \qquad F_f^1 = H_{fr} F_r,$$

$$G_r^1 = H_{rr} G_r, \qquad G_f^1 = H_{fr} G_r,$$

$$B_r^1 = H_{rr} B_r + H_{rf} B_f, \qquad B_f^1 = H_{fr} B_r + H_{ff} B_f.$$

The expression given by (9.3.3) is more suitable to apply to singular perturbation theory.

9.3.2   Singular perturbation approach

The singular perturbation approach (Kokotović (1984)) consists of breaking the dynamics of the system into two parts, each of them in a separate time scale. The slow dynamics corresponds to the rigid modes q_r and the fast dynamics to the flexible modes q_f. Introduce a small scale factor ε and define

$$\varepsilon^2 \xi = q_f, \qquad \tilde{K}_{ff} = \varepsilon^2 K_{ff}, \tag{9.3.5}$$

where $\frac{1}{\varepsilon^2}$ is equal to the smallest stiffness in K_{ff}.

Define the control u as

$$u = \bar{u} + u_F \tag{9.3.6}$$

with $\bar{u}$ and u_F being the slow and fast control components, respectively.

Substituting (9.3.5) and (9.3.6) in (9.3.3) and setting $\varepsilon = 0$, we can find the slow manifold to be

$$\bar{\xi} = \tilde{K}_{ff}^{-1} \bar{H}_{ff}^{-1} (-\bar{V}_{fr}^1 \dot{\bar{q}}_r - \bar{F}_f^1 - \bar{G}_f^1 + \bar{B}_f^1 \bar{u}), \tag{9.3.7}$$

where the overbar indicates the evaluation of nonlinear functions at $\varepsilon = 0$. The slow subsystem, which is the rigid model, is given by

$$\ddot{\bar{q}}_r = \bar{M}_{rr}^{-1} [-\bar{V}_{rr} \dot{\bar{q}}_r - \bar{F}_r - \bar{G}_r + \bar{B}_r \bar{u}] \tag{9.3.8}$$

and is obtained using (9.3.3), (9.3.5), and (9.3.7) and setting $\varepsilon = 0$.

Define $\varsigma_1 \equiv \xi - \bar{\xi}$ and $\varsigma_2 \equiv \varepsilon \dot{\xi}$. Using (9.3.3), (9.3.5), (9.3.7), and a time-scale change of $T = \frac{t}{\varepsilon}$, the fast dynamics can be found to be

$$\frac{d}{dT} \begin{bmatrix} \varsigma_1 \\ \varsigma_2 \end{bmatrix} = \begin{bmatrix} 0 & I \\ -\bar{H}_{ff} \bar{K}_{ff} & 0 \end{bmatrix} \begin{bmatrix} \varsigma_1 \\ \varsigma_2 \end{bmatrix} + \begin{bmatrix} 0 \\ \bar{B}_f^1 \end{bmatrix} u_F = A_F \varsigma + B_F u_F. \tag{9.3.9}$$

According to Tikhonov's theorem, the flexible-link robot can be described to order ε using (9.3.8) and (9.3.9). In fact, as long as we apply the control given by (9.3.6), the trajectories in the original system are given by

$$q_r = \bar{q}_r + O(\varepsilon),$$
$$q_f = \varepsilon^2 (\bar{\xi} + \varsigma_1) + O(\varepsilon), \tag{9.3.10}$$

with $O(\varepsilon)$ denoting terms of order ε.

Define the modified tracking output as

$$y \equiv \begin{bmatrix} \bar{q}_r \\ \dot{\bar{q}}_r \end{bmatrix}.$$ (9.3.11)

Note that we are only concerned to make the slow part of the tip position track the desired trajectory. This leaves some freedom in selecting the control input which is going to be used to damp out the vibratory fast modes.

The internal dynamics relative to $y(t)$ of this modified tracking output are given by the fast subsystem described in (9.3.9). The fast control component u_F can be chosen to give stable internal dynamics if the pair (A_F, B_F) is stabilizable, which is generally the case for flexible-link arms. Under this assumption, there are many approaches in robust linear system theory that yield a stabilizing control. The fast control is of the form

$$u_F = -[K_{pF} \quad K_{dF}] \cdot \begin{bmatrix} \varsigma_1 \\ \varsigma_2 \end{bmatrix} = -\frac{K_{pF}}{\varepsilon^2} q_f - \frac{K_{dF}}{\varepsilon} \dot{q}_f + K_{pF}\bar{\xi}.$$ (9.3.12)

9.3.3 Neural network control algorithm

The slow dynamics of a flexible-link robot arm is given by

$$\bar{M}_{rr} \cdot \ddot{\bar{q}}_r + \bar{D}_{rr} \cdot \dot{\bar{q}}_r + \bar{F}_r + \bar{G}_r = \bar{B}_r \bar{u},$$ (9.3.13)

which is exactly the Lagrange form of an n-link rigid robot arm, satisfying the standard rigid robot properties (Lewis, Abdallah, and Dawson (1993)). For this dynamics, an NN controller can be designed.

Given a desired trajectory $q_d(t)$ for q_r the tracking error is defined as

$$e = q_d - \bar{q}_r$$ (9.3.14)

and the filtered tracking error as

$$r = \dot{e} + \Lambda \cdot e,$$ (9.3.15)

where $\Lambda = \Lambda^T > 0$. Using (9.3.15), the slow dynamics in (9.3.13) can be rewritten in terms of the filtered tracking error as

$$\bar{M}_{rr} \cdot \dot{r} = -\bar{D}_{rr} \cdot r - \bar{B}_r \bar{u} + f(x),$$ (9.3.16)

where the nonlinear robot function is

$$f(x) = \bar{M}_{rr}(\bar{q}_r) \cdot (\ddot{\bar{q}}_d + \Lambda \cdot \dot{e}) + \bar{D}_{rr}(\bar{q}_r, \dot{\bar{q}}_r) \cdot (\dot{\bar{q}}_d + \Lambda \cdot e) + \bar{F}_r(\dot{\bar{q}}_r) + \bar{G}_r(\bar{q}_r)$$ (9.3.17)

with

$$x = [e \quad \dot{e} \quad q_d \quad \dot{q}_d \quad \ddot{q}_d]^T.$$ (9.3.18)

The unknown function $f(x)$ will be estimated using a two-layer NN exactly as in previous chapters. Thus we assume that

$$f(x) = W^T \sigma(V^T x) + \varepsilon$$ (9.3.19)

with the NN approximation error $\varepsilon(x)$ bounded on a compact set. The estimate of $f(x)$ is given by

$$\hat{f}(x) = \hat{W} \cdot \sigma(\hat{V}^T \cdot x) \tag{9.3.20}$$

with hats denoting the actual NN weight values in the system. Let the ideal unknown weight matrix be given by

$$Z = \begin{bmatrix} V & 0 \\ 0 & W \end{bmatrix}. \tag{9.3.21}$$

It is assumed bounded by $\|Z\| \le Z_m$ with Z_m a known bound.

It is assumed that the desired trajectory is bounded according to

$$\left\| \begin{matrix} q_d \\ \dot{q}_d \\ \ddot{q}_d \end{matrix} \right\| \le Q \tag{9.3.22}$$

with Q a known bound.

In Lewis, Jagannathan, and Yesildirek (1999) it is shown that the closed-loop system is UUB if one selects the rigid system control law

$$\bar{u} = \bar{B}_r^{-1} \hat{W}^T \sigma(\hat{V}^T x) + K_r r - v \tag{9.3.23}$$

with the robustifying term given by

$$v(t) = -K_z(\|\hat{Z}\| + Z_m)r(t) \tag{9.3.24}$$

with gain K_z selected large enough. The NN weights are tuned using the algorithms

$$\begin{aligned}
\dot{\hat{W}} &= F \cdot \hat{\sigma} \cdot r^T - F \cdot \hat{\sigma}' \cdot \hat{V}^T \cdot x \cdot r^T - \kappa \cdot F \cdot \|r\| \cdot \hat{W}, \\
\dot{\hat{V}} &= G \cdot x \cdot (\hat{\sigma}'^T \cdot \hat{W} \cdot r)^T - \kappa \cdot G \cdot \|r\| \cdot \hat{V}
\end{aligned} \tag{9.3.25}$$

with any constant matrices $F = F^T > 0$ and $G = G^T > 0$, and scalar tuning parameter $\kappa > 0$. These are modified backprop tuning laws that guarantee stability and avoid the need for a "persistence of excitation" (PE). The first terms in each algorithm are standard backpropagation. The last terms are equivalent to Narendra's e-mod used in adaptive control; these terms keep the weight estimation errors bounded.

Using the constructions given above, the filtered tracking error $r(t)$ and the NN weight errors $\tilde{V}$ and $\tilde{W}$ are UUB. Moreover, the tracking error may be kept as small as desired by increasing the gains K_p and K_v.

9.4 Implementation of neural network controller on ATB1000 testbed

Under a U.S. Army SBIR contract with Simis Labs, Inc., we implemented the NN flexible-link pointing controller just developed on the ATB1000 testbed in Picatinny, NJ. The implementation was done using the real-time PC–PC controller developed at the ARRI. Figure 9.4.1 shows the experimental setup used in this implementation. The NN controller

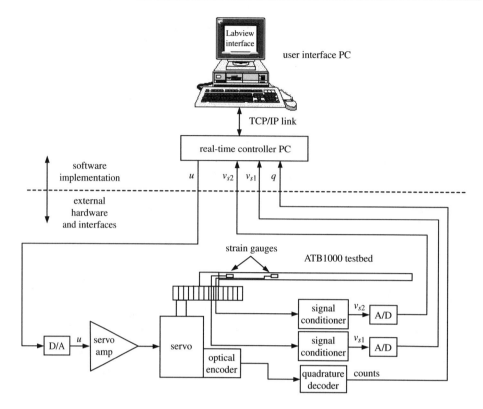

Figure 9.4.1. *Experimental setup for ATB1000 NN controller implementation.*

was written in C for the RTCS target computer; the controller code is given in Appendix D of the book.

Only the first two flexible modes for the barrel were considered. To measure the amplitudes of these two modes, we installed strain gauges on the ATB1000 barrel. Since we were trying to control both azimuth and elevation, we required two strain gauges on the top of the barrel and two on the side. With two strain gauges in each dimension, one can use standard Mechanical Engineering tools to compute the amplitudes of the first two modes in both azimuth and elevation. One takes an infinite series expansion for the solution to the Bernoulli–Euler beam equation, truncates the series, and uses assumed mode shape techniques.

In this chapter, we show only the implementation results for the azimuth degree of freedom. The sampling period used in the RTCS was selected as 2 ms. The power of RTCS is shown by the fact that we can perform all computations needed to implement the NN controller in 2 ms. This would not be possible using a standard controller board available commercially.

The reference input was selected as a 0.2Hz square wave input with 30 degrees amplitude. The square wave gives alternating step commands in each direction that simulate the aiming and shooting of different targets located on both sides of the tank. The square input was passed through a trajectory generator to take the jerk out of the movements.

To perform a comparative performance study, two different experiments were conducted: one using a simple PD control feedback, and another one using the singular perturbation NN controller derived in this chapter.

9.4.1 Proportional-plus-derivative control

First, to provide a baseline performance comparison, a simple PD controller was implemented. That is, we used the control law given by (9.3.6) with the PD slow control

$$\bar{u} = K_p \cdot e + K_v \cdot \dot{e} \qquad (9.4.1)$$

and the fast control

$$u_F = -[K_{pF} \quad K_{vF}] \cdot \begin{bmatrix} q_f \\ \dot{q}_f \end{bmatrix}. \qquad (9.4.2)$$

The control gains were selected as $K_p = 0.35$, $K_v = 0.04$, $K_{pF} = [0.5 \quad 0.1]$, and $K_{vF} = [0.01 \quad 0.01]$. These gains were selected after several implementation trials with different gains.

Figure 9.4.2 shows the reference position and the actual tip azimuth position measured in degrees. It can be seen that the magnitude of the steady-state error is about 2 degrees. This error will result in missing the target by about 61 yards for a one-mile target range. This is not accurate enough for most battlefield applications.

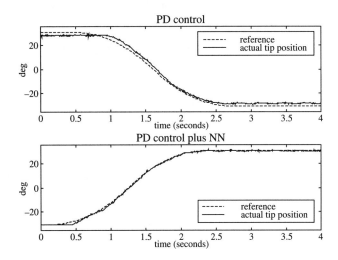

Figure 9.4.2. *Results of ATB1000 barrel aiming implementation.*

9.4.2 Proportional-plus-derivative control plus neural network

Now the singular perturbation NN controller developed in this chapter was implemented on the ATB1000. The NN has 10 neurons in the hidden layer, one output, and the input vector is given by (9.3.18). The NN tracking controller was implemented using the parameters $F = 0.15$, $G = 0.01$, and $\kappa = 2$ and the same control gains used for the PD rigid control experiment. To select these values, we ran several implementation trials with different values of the design parameters to obtain this set, which gave good results.

A control run of one minute was conducted. At this time, the azimuth angle error dropped by a significant amount. It was difficult to speed up the learning because the ATB1000 has an adjustable backlash mechanism, and the mechanical resonances of this backlash mechanism were excited when training the NN too fast. An NN backlash controller was given in Lewis, Jagannathan, and Yesildirek (1999). When the backlash is compensated for, the training can be reduced to only a few seconds, as shown in work on a different testbed. These are the sorts of things that one discovers and must confront when performing implementations on actual hardware.

Figure 9.4.2 shows the reference position and the actual tip azimuth position measured in degrees. It can be seen that the magnitude of the steady-state error is negligible (less than 0.05 degrees). This results in extreme accuracy of firing of the tank canon. Figure 9.4.3 depicts the NN output and the total control signal. It can be seen the NN output signal has a final offset to take care of the steady-state error.

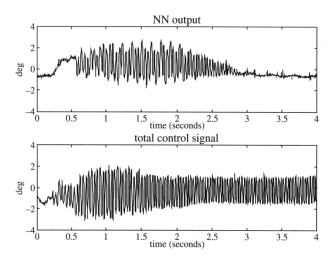

Figure 9.4.3. *NN and total control signal outputs.*

Note that the step command does not occur exactly at the same time for the two cases studied, as can be seen in Figure 9.4.2. This is mainly because the implementation was done using a real-time controller. The timing of the recorded results depends on when the signal can be captured by the RTCS for display and recording.

9.5   Conclusions

The practical implementation of a multiloop nonlinear NN tracking controller for a single flexible link has been tested, and its performance compared to the one of standard PD control. The controller includes an outer PD tracking loop, a singular perturbation network inner loop for stabilization of the fast dynamics, and an NN inner loop used to feedback-linearize the slow dynamics. This NN controller requires no offline learning phase (the NN weights are easily initialized) and guarantees boundedness of the tracking error and control signal.

The NN controller was implemented using the RTCS controller developed at UTA's ARRI. RTCS allows implementation of complex, advanced, nonlinear control algorithms on actual industrial and Department of Defense systems. The practical results show the improvement achieved in tracking using the NN. The NN is able to learn the unmodeled dynamics of the system to compensate for the barrel flexible modes and yield accurate pointing.

Appendix A

C Code for Neural Network Friction Controller

This appendix contains the C code for the NN friction controller simulation in Chapter 3. For more information, see Selmic (2000).

```
/*************************************************************
This is a simulation of serial robot manipulator system with
friction using augmented neural network.

File name: friction.c
Authors: Young Ho Kim and Rastko R. Selmic

Date: May 26, 1997

Matrices are defined as following

   V = v01 v02....v0L          W = w01 w02....w0m
     v11 v12....v1L            w11 w12....w1m
       .   .    .                .   .    .
       .   .    .                .   .    .
     vn1 vn2 vnL [(n+1), L]      wL1 wL2...wLm [(L+1), m]

Matrices are stored in column fashion, so first is stored the
first column, then second, and so on.

   weight_v[0] = v01     weight_w[0] = w01
   weight_v[1] = v11     weight_w[1] = w11
         .                     .
         .                     .
   weight_v[n] = vn1     weight_w[L] = wL1
   weight_v[n+1] = v02   weight_w[L+1] = w02
         .                     .
         .                     .

*************************************************************/

#include <stdio.h>
```

```c
#include <math.h>
#include <stdlib.h>
#include <malloc.h>

#define pi M_PI
#define time_step 0.0005
#define T 10
#define print_step 100
#define n_joint 2  // number of robot joints
#define n_input 10  // NN inputs
#define n_hidden 10 // number of hidden nodes
#define n_output 2
// NN output, should be equal to number of robot joints

#define n_hidden_augment 4
// number of augmented hidden layer nodes

void Create_desired();
// function to create the desired trajectory

void Compute_traj_filter_error();
// function to compute the filter tracking error

void Compute_robust_term();
// function to compute the robustifying term

// Runge-Kutta (4. order) for integrating the differential
   equations
void rk4(void (*rastko)(double x[],double tau[],
 double xdot[]),
int order, double x_in[],double u[],double x_out[],double t0,
 double tf);

// robot dynamics
void robot_dyn(double x[],double tau[],double xdot[]);

void weight_v_dyn(double x[],double tau[],double xdot[]);

void weight_w_dyn(double x[],double tau[],double xdot[]);

// dynamics of the augmented NN
void weight_w_augment_dyn(double x[],double tau[],
 double xdot[]);

void Initialize_sate_weight();

// n_interval can be n_hidden for alpha gain,
 (n_input+1)*n_hidden for first layer
void Uniform(double a, double b, int n_interval,
```

```
 double average[]);

void Input_neural();

void Weight_update_v();

void Weight_update_w();

void Hidden_output();

void Neural_output();

double Lambda = 5.0, Kv=20.0; // PD gain
double Kz = -2.0, Kz1 = -15.0, W_max = 5;
// robustifying gains
double t = 0.0;
double kk = 0.01;  // NN learning for e or sigma mod
double F = 10.0, G = 10.0, E = 35.0; // NN learning gains

int fNN_include = 1;  // = 1 NN compensator included
   // = 0 without NN compensator
int fFriction_include = 1;  // = 1 friction model included
   // = 0 without friction model
int fAugmented_NN = 1;  // = 1 augmented NN included
   // = 0 without augmented NN

//Declare the matrix storage for robot and neural net system
double qd[n_joint], qdp[n_joint], qdpp[n_joint];
double robot_x[2* n_joint];
double e[n_joint], ep[n_joint];
double r[n_joint+1];
double norm_r;
double v_robust[n_joint];
double tau[n_joint];
double norm_weight;

// Define matrix for neural network structure, weight and
   sigmoid
double input_x[n_input+1];
double hidden_net[n_hidden+1], hidden_output[n_hidden+1];
double alpha[n_hidden+1];
double neural_output[n_output];
double weight_v[(n_input+1)*n_hidden];
double weight_w[(n_hidden+1)*n_output];
double tau_weight_v[(n_input+1)*n_hidden],
 tau_weight_w[(n_hidden+1)*n_output];

// Arrays used for the augmented NN
double hidden_output_augment[n_hidden_augment+1];
```

```c
double weight_w_augment[(n_hidden_augment+1)*n_output];
double tau_weight_w_augment[(n_hidden_augment+1)*n_output];

int main() {

 int i, k=10;
 FILE *frobot_x, *fwei_v, *fwei_w;

 // Open files for storage of the data
 frobot_x=fopen("c:\\rs\\csim\\x_robot","w");
 fwei_v=fopen("c:\\rs\\csim\\x_wei_v","w");
 fwei_w=fopen("c:\\rs\\csim\\x_wei_w","w");

 // Print to file only to tell Matlab which option is
    simulated
 fprintf(frobot_x, "%d %d %d %d %d %d %d %d %d %d %d \n",
 fNN_include,
    fFriction_include, fAugmented_NN, k, k, k, k, k, k, k, k);

 // Initialize the robot states,weight values,and randomize
    alpha gain.
 Initialize_sate_weight();

 t=0.0;
 do {
  Create_desired();  // Create the desired trajectory

  Compute_traj_filter_error();

  // Calculate the NN for nonlinear function approximation
  if(fNN_include == 1){
   Input_neural();
   // Prepare the input data to neural network
   Hidden_output();
   // Using current or initial state and weight values
   Weight_update_v();
   // Update weights based on the current system state or
      initial system state
   Weight_update_w();
   Hidden_output();
   // Using the updated weight values and current states
   Neural_output();
   // Using the updated weight values and current states
  }
  // Compute_robust_term();

  // Control input to the robot system
  for(i=0; i<n_joint; i++)
   tau[i] = r[i+1] * Kv + neural_output[i] - 0.0
```

```
  * v_robust[i];

 // Print the current values
 if(k%print_step==0){
  k=0;

  printf("%f\n",t);

  fprintf(frobot_x, "%f %f %f %f %f %f %f %f %f %f %f\n",
  t, e[0], e[1],
   ep[0], ep[1], qd[0], qd[1], neural_output[0],
   neural_output[1], r[1], r[2]);

  fprintf(fwei_v,"%f %f\n",weight_v[1],weight_v[n_input+1]);

  fprintf(fwei_w, "%f %f\n",weight_w[1],weight_w[2]);
 }
 k++;

 // Runge - Kutta of 4. order
 rk4(robot_dyn, 2*n_joint, robot_x, tau, robot_x, t,
 t+time_step);

  t += time_step;
 } while (t<=T);

 fcloseall();

 return 0;
}

void robot_dyn(double x[],double tau[],double xdot[]) {
/***************************************************************
 Simulation of the equation
 xdot = -ax + tau

 INPUT VALUES: x, tau

 RETURNED VALUE: xdot
 ***************************************************************/

 // Parameters of the robot dynamics
 double m11, m12, m22, v1, v2, a1, a2, g1, g2, det;
 double m1=3.0, m2=2.3, le1=1.1, le2=1.0, g=9.8;
 double frict1, frict2;

 // Friction model parameters for the first joint
 double a0_frict = 35, a1_frict = 1.1, a2_frict = 0.9;
 double b1_frict = 50, b2_frict = 65;
```

```
// Friction model parameters for the second joint:
double c0_frict = 38, c1_frict = 1, c2_frict = 0.95;
double d1_frict = 55, d2_frict = 60;

// Robot dynamics
m11 = (m1+m2)*le1*le1 + m2*le2*le2 + 2.*m2*le1*le2*cos(x[1]);
m22 = m2*le2*le2;
m12 = m22 + m2*le1*le2*cos(x[1]);
v1 = -m2*le1*le2*(2.*x[2]*x[3] + x[3]*x[3])*sin(x[1]);
v2 = m2*le1*le2*x[2]*x[2]*sin(x[1]);
g1 = (m1+m2)*g*le1*cos(x[0]) + m2*g*le2*cos(x[0] + x[1]);
g2 = m2*g*le2*cos(x[0] + x[1]);
det = m11*m22 - m12*m12;

if (fFriction_include = 1) {
 // Friction force for the 1. joint
 if (x[2] > 0)
  frict1 = a0_frict+a1_frict*exp(-b1_frict*x[2])
  +a2_frict*(1-exp(-b2_frict*x[2]));
 else
  frict1 = -(a0_frict+a1_frict*exp(b1_frict*x[2])
  +a2_frict*(1-exp(b2_frict*x[2])));

 // Friction force for the 2. joint
 if (x[3]>0)
  frict2 = c0_frict+c1_frict*exp(-d1_frict*x[3])
  +c2_frict*(1-exp(-d2_frict*x[3]));
 else
  frict2 = -(c0_frict+c1_frict*exp(d1_frict*x[3])
  +c2_frict*(1-exp(d2_frict*x[3])));

 a1 = tau[0] - v1 - g1 - frict1;
 a2 = tau[1] - v2 - g2 - frict2;
 }
 else {
 a1 = tau[0] - v1 - g1;
 a2 = tau[1] - v2 - g2;
 }
 // State space model of the robot
 xdot[0] = x[2];
 xdot[1] = x[3];
 xdot[2] = ( m22*a1 - m12*a2 )/det;
 xdot[3] = ( m11*a2 - m12*a1 )/det;
}

void Input_neural() {

 // NN input
 input_x[0] = 1.0;   // Update with 1 for a threshold
```

```
  input_x[1]  = e[0];
  input_x[2]  = e[1];
  input_x[3]  = qd[0];
  input_x[4]  = qd[1];
  input_x[5]  = ep[0];
  input_x[6]  = ep[1];
  input_x[7]  = qdpp[0];
  input_x[8]  = qdpp[1];
  input_x[9]  = qdp[0];
  input_x[10] = qdp[1];
}

void Hidden_output(){
  // Calculating the output value of the hidden layer

  int i, j, offset;
  double net;

  for(j=1; j<=n_hidden; j++) {
   net=0.0;
   for(i=0; i<=n_input; i++) {
    offset=(j-1)*(n_input+1)+i;
    net += input_x[i]*weight_v[offset];
   }
   hidden_output[j]=1./(1+exp(-net));
   hidden_net[j]=net;
  }
  hidden_net[0] = 1;
  hidden_output[0] = 1;

  if (fAugmented_NN == 1) {
   //Generation of the output of the augmented hidden neurons
   if (robot_x[2]>0) {
    hidden_output_augment[1] = 1.0;
    hidden_output_augment[2] = 1.0 - exp(-robot_x[2]);
   }
   else {
    hidden_output_augment[1] = 0.0;
    hidden_output_augment[2] = 0.0;
   }
   if (robot_x[3]>0) {
    hidden_output_augment[3] = 1.0;
    hidden_output_augment[4] = 1.0 - exp(-robot_x[3]);
   }
   else {
    hidden_output_augment[3] = 0.0;
    hidden_output_augment[4] = 0.0;
   }
   hidden_output_augment[0] = 1.0;
  }
}
```

```
void Neural_output() {
 // Calculating the output value of the output layer

 int i,j,offset;

 for(j=0; j<n_output; j++) {
  neural_output[j] = 0.0;
  for(i=0; i<=n_hidden; i++){
   offset = j*(n_hidden+1) + i;
   neural_output[j] += hidden_output[i] * weight_w[offset];
  } //NN activation function is linear for the output layer
  // If there is augmented NN, then add that part into the
     NN output
  if (fAugmented_NN == 1) {
   for(i=0; i<=n_hidden_augment; i++){
    offset = j*(n_hidden_augment+1) + i;
    neural_output[j] += hidden_output_augment[i]
     * weight_w_augment[offset];
   }
  }
 }
}


void Weight_update_v() {
 // Function for updating the V weights

 int ii, j, k, offset;
 double sigp, alpha;

 // Prepare the control input to the first-layer weight
    update
 for(j=1; j<=n_hidden; j++){
  alpha = 0.0;
  offset = j;
  sigp=(1 - hidden_output[j])* hidden_output[j];
  for(ii=1; ii<=n_output; ii++) {
   alpha += weight_w[offset]*r[ii];
   offset += n_hidden+1;
  }
  for(k=0; k<=n_input; k++) {
   offset=(j-1)*(n_input+1)+k;
   tau_weight_v[offset]= G * sigp * alpha * input_x[k];
  }
 }
 rk4(weight_v_dyn, (n_input+1)*n_hidden, weight_v,
 tau_weight_v, weight_v, t, t+time_step);
}

void weight_v_dyn(double xx[],double tau[], double xdot[]) {
```

```
/***********************************************************
 Dynamics of the NN weights

 INPUT VALUES: xx = V(hat)
     tau = G*(sigmaprime(transpose) * W *r) (transpose)

 RETURNED VALUE: xdot = V hat)
 ***********************************************************/

 int i, j, offset;

 for(i=1; i<=n_hidden; i++) {
  for(j=0; j<=n_input; j++) {
   offset=(i-1)*(n_input+1)+j;
   xdot[offset]=-kk*G*norm_r*xx[offset]+tau[offset];
  }
 }
}

void Weight_update_w() {
 // Function for updating the W weights

 int i,j,offset;
 double sigp;

 // Prepare the control input to the second layer weight
    updater
 for(j=1; j<=n_output; j++) {
  for(i=0; i<=n_hidden; i++){
   offset=(j-1)*(n_hidden+1) + i;
   sigp=(1- hidden_output[i])* hidden_output[i];
   tau_weight_w[offset]=F*hidden_output[i]*r[j]
   -F*sigp*hidden_net[i]*r[j];
  }
 }
 // Integrate the diff. equation
 rk4(weight_w_dyn, (n_hidden+1)*n_output, weight_w,
 tau_weight_w, weight_w, t, t+time_step);
 // If augmented NN is used, then update the output augmented
    weights
 if (fAugmented_NN == 1) {
  for(j=1; j<=n_output; j++) {
   for(i=0; i<=n_hidden_augment; i++){
    offset = (j-1)*(n_hidden_augment+1) + i;
    tau_weight_w_augment[offset]
    = E*hidden_output_augment[i]*r[j];
   }
  }
  // Integrate the diff. equation
```

```
  rk4(weight_w_augment_dyn, (n_hidden_augment+1)*n_output,
  weight_w_augment, tau_weight_w_augment,
weight_w_augment, t, t+time_step);
 }
}

void weight_w_dyn(double xx[],double tau[], double xdot[]) {
/************************************************************
 Dynamics of the NN weights

 INPUT VALUES: xx = W (hat)
     tau = F*sigma*r(transpose)
     - F*sigma prime*V(transp)*x*r(transp)

 RETURNED VALUE: xdot = W (hat)
 *************************************************************/

 int i,j,offset;
 for(i=1; i<=n_output; i++) {
  for(j=0; j<=n_hidden; j++){
   offset=(i-1)*(n_hidden+1) + j;
   xdot[offset] = -kk*F*norm_r*xx[offset]+tau[offset];
  }
 }
}

void weight_w_augment_dyn(double xx[],double tau[], double
xdot[]) {
/************************************************************
 Dynamics of the NN weights

 INPUT VALUES: xx = W2 (hat)
     tau = E*fi*r(transpose)

 RETURNED VALUE: xdot = W prime (hat)
 *************************************************************/

 int i,j,offset;
 for(i=1; i<=n_output; i++) {

  for(j=0; j<=n_hidden_augment; j++){
   offset=(i-1)*(n_hidden_augment+1) + j;
   xdot[offset] = -kk*E*norm_r*xx[offset]+tau[offset];
  }
 }
}

void Create_desired() {
 // Calculating the desired trajectory
```

```
double amp = 1.0;
double w = 1.0;


qd[0] = amp * sin(w*t);
// Desired position for the 1-joint
qd[1] = amp * cos(w*t);
// Desired position for the  2-joint

qdp[0] = amp * w * cos(w*t);
// Desired velocity for the 1-joint
qdp[1] = -amp * w * sin(w*t);
// Desired velocity for the 2-joint

qdpp[0] = -amp * w * w * sin(w*t);
// Desired acceleration for the 1-joint
qdpp[1] = -amp * w * w * cos(w*t);
// Desired acceleration for the 2-joint
}

void Compute_traj_filter_error() {

  int i;
  norm_r=0;

  for(i=0; i<n_joint; i++) {
   e[i] = qd[i] - robot_x[i]; // Error
   ep[i] = qdp[i] - robot_x[i+2];
   // Error prime (derivative of the error)
   r[i+1] = ep[i] + Lambda*e[i]; // r = ep + Lambda * e
   norm_r += r[i+1] * r[i+1]; // Norm of the error
  }
  norm_r = sqrt(norm_r);
}

void Compute_robust_term() {

  int i;

  // Calculate the robustifying terms
  for(i=0; i<n_joint; i++) {
   if (i==0) Kz=Kz1;
   v_robust[i]=-Kz*(norm_weight+W_max)*r[i+1];
  }
}

void Initialize_sate_weight() {

  int i;
```

```
double a=0.5,b=1.5,a_v=-1,b_v=1;
double *average;
// reserved for return value from uniform random generator
for(i=0; i<2*n_joint; i++) robot_x[i]=0.0;
for(i=0; i<(n_input+1)*n_hidden; i++) {
 weight_v[i] = a_v+(b_v-a_v) * rand()/RAND_MAX;
}
for(i=0; i<(n_hidden+1)*n_output; i++) {
 weight_w[i]=0.;
 printf("%f\n", weight_w[i]);
}
average=calloc(n_hidden+1, sizeof(double));
Uniform(a,b,n_hidden,average);
for(i=1;i<=n_hidden;i++) {
 alpha[i]=1.0;
 printf("%f\n", alpha[i]);
}
free(average);
}


void Uniform(double a, double b, int n_interval, double
average[]) {

 int i, j, n=100;
 // n is the times of generating random number
 double random_number, _uniform_data, delta;
 double *sum_interval, *no_interval;
 // no_interval:the number of random number inside a certain
    interval

 delta=(b-a)/n_interval;
 sum_interval = calloc(n_interval+1, sizeof(double));
 no_interval = calloc(n_interval+1, sizeof(double));
 for(j=1; j<=n_interval; j++) sum_interval[j]=no_interval[j]
=0;
 for(i=1;i<=n;i++){
  random_number = (double)rand()/RAND_MAX;
  _uniform_data = a+(b-a)*random_number;
  for(j=1;j<=n_interval;j++){
   if(_uniform_data >=a+(j-1)*delta && _uniform_data
   < a+j*delta){
     sum_interval[j]+=_uniform_data;
     no_interval[j]++;
   }
  }
 }
 for(j=1;j<=n_interval;j++){
  average[j]=sum_interval[j]/no_interval[j];
 }
```

```
free(sum_interval);free(no_interval);
}

void rk4(void (*rastko)(double x[], double tau[],
double xdot[]),
          int order, double x_in[], double u[],
          double x_out[], double t0, double tf) {
/************************************************************
Runge Kutta 4. order

Runge Kutta for integrating the differential equations
Usage:
  1. file which contains the dynamics of the system,
  2. number of the states
  3. x_in - input states
  4. u - additional input
  5. x_out - output states after the step of integration
  6. t0 - starting time
  7. tf - final time
************************************************************/

  int i;
  double h = tf-t0;
  double h2 = h/2.0;
  double h3 = h/3.0;
  double h6 = h/6.0;
  double *x_mid1,*x_mid2,*x_end,*xdot_init,*xdot_mid1;
  double *xdot_mid2,*xdot_end;

  x_mid1 = calloc(order, sizeof(double));
  x_mid2 = calloc(order, sizeof(double));
  x_end = calloc(order, sizeof(double));
  xdot_init = calloc(order, sizeof(double));
  xdot_mid1 = calloc(order, sizeof(double));
  xdot_mid2 = calloc(order, sizeof(double));
  xdot_end = calloc(order, sizeof(double));

  (*rastko)(x_in, u, xdot_init); // get xdot at initial x

  for (i=0; i<order; i++) x_mid1[i] = x_in[i]
  + h2*xdot_init[i];
  (*rastko)(x_mid1, u, xdot_mid1);
  // get xdot at first guess  midpt
  for (i=0; i<order; i++) x_mid2[i] = x_in[i]
  + h2*xdot_mid1[i];
  (*rastko)(x_mid2, u, xdot_mid2);
  // get xdot at second guess midpt
  for (i=0; i<order; i++) x_end[i] = x_in[i] + h*xdot_mid2[i];
  (*rastko)(x_end, u, xdot_end);
```

```
// get xdot at estimated x(t+dt)
for (i=0; i<order; i++)
 x_out[i] = x_in[i] + h6*(xdot_init[i]+xdot_end[i])
     + h3*(xdot_mid1[i]+xdot_mid2[i]);
free(x_mid1); free(x_mid2); free(x_end); free(xdot_init);
free(xdot_mid1); free(xdot_mid2); free(xdot_end);
}
```

Appendix B

C Code for Continuous-Time Neural Network Deadzone Controller

This appendix contains the C code for the continuous-time NN deadzone controller simulation in Chapter 4. For more information, see Selmic and Lewis (2000).

```
/*************************************************************
This is control simulation of serial robot manipulator
system with deadzone

File name: deadz.c
Author: Rastko R. Selmic

Date: July 1998

Matrices are defined in the following way

  V = v01 v02....v0L    W = w01 w02....w0m
    v11 v12....v1L        w11 w12....w1m
     .   .   .             .   .   .
     .   .   .             .   .   .
    vn1 vn2 vnL [(n+1), L]    wL1 wL2...wLm [(L+1), m]
*************************************************************/

#include <stdio.h>
#include <math.h>
#include <stdlib.h>
#include <malloc.h>

// Parameters for simulation, time step, plotting time
#define time_step 0.001
#define print_step 50
#define tfinal 12

// Parameters of the NN I (neural network used as the
   estimator)
#define N1 2 // Input layer
```

```
#define N2 20 // Hidden layer
#define N3 2 // Output layer

// Parameters of the NN II (neural network used as
   compensator)
#define N1i 2 // Input layer
#define N2i 20 // Hidden layer
#define N3i 2 // Output layer
#define Nai 4 // Jump layer

#define n 4 // number of system states

void rk4(void (*rastko)(double* x, double* tau, double* xdot),
     int order, double* x_in, double* u,
     double* x_out, double t0, double tf);

void testdyn(double* x, double* tau, double* xdot);

// Function for calculating the xdot based on robot dynamics
void robot_dyn(double* x, double* tau, double* xdot);

// Function for NN W-weights tuning law;
void weights_W(double* W, double* tau, double* Wdot);

// Function for NN Wi-weights tuning law;
void weights_Wi(double* Wi, double* tau, double* Widot);

// Transpose of matrix
void MatrixTransp (double *A1, int m, int n, double *R,
int p, int q);

// Sum of 2 matrices
void MatrixAdd (double *A1, int m1, int n1,
        double *A2, int m2, int n2,
        double *R, int p, int q);

// Multiplication of the matrices
void MatrixMul (double *A1, int m1, int n1,
        double *A2, int m2, int n2,
        double *R, int p, int q);

// Multiplication of 3 matrices
void MatrixMul3 (double *A1, int m1, int n1,
        double *A2, int m2, int n2,
        double *A3, int m3, int n3,
        double *R, int p, int q);

// Multiplication of 4 matrices
void MatrixMul4 (double *A1, int m1, int n1,
```

```
          double *A2, int m2, int n2,
          double *A3, int m3, int n3,
          double *A4, int m4, int n4,
          double *R, int p, int q);

// Multiplication of 5 matrices
void MatrixMul5 (double *A1, int m1, int n1,
          double *A2, int m2, int n2,
          double *A3, int m3, int n3,
          double *A4, int m4, int n4,
          double *A5, int m5, int n5,
          double *R, int p, int q);

double t = 0.0;
double tuning_S = 240.0; // tuning parameter for NN I
double tuning_T = 500.0; // tuning parameter for NN II
double k1 = 0.001; // parameters for tuning law
double k2 = 0.0001;

// External variables which are used by NN tuning law
double norm_r = 0; // norm of the filtered tracking error

int N2_nn = N2;
// NN2_nn and NN3_nn are just for external variables
int N3_nn = N3;
int N2i_nn = N2i;
int N3i_nn = N3i;
int Nai_nn = Nai;

double x[n]; // states of the system;
double V[N1][N2], v0[N2]; // NN I
double V_min, V_max, v0_min, v0_max;
// Initialization for NN I
double Vi[N1i][N2i], v0i[N2i]; // NN II
double W_min, W_max; // Initialization for NN II
double W[N2+1][N3]; // NN I second layer weights;
double W1i[N2i+1][N3i],W2i[Nai+1][N3i];
// NN II second layer and augmented layer weights
double Wi[N2i+Nai+2][N3i]; // NN II second layer all weights
double fi[Nai+1];
double sigmai[Nai+N2i+2][1];
double NNI[N3][1]; // NNII output;
double NNII[N3i][1]; // NNII output;

double RAN_MAX = 32676.0;

double qd[2], qdp[2], qdpp[2];
// desired position, velocity and acceleration
double e[2], edot[2]; // tracking errors;
```

```c
double r[2]; // filtered tracking errors;
double w[2], u[2], tau[2]; // signals w, u, and tau;
double Lambda[2][2]; // parameter Lambda;
double Kv; // PD gain;

int main() {

 int i, j, k;

 srand(10000);

 // PARAMETERS in simulation
 Kv = 20.0;
 Lambda[0][0] = 7.0; Lambda[0][1] = 0.0;
 Lambda[1][0] = 0.0; Lambda[1][1] = 7.0;

 // Deadzone parameters
 double d_plus = 25.0; double d_minus = -20.0;
 double m_plus = 1.5; double m_minus = 0.8;

 // Random numbers are uniformly distributed between v_min
    and v_max
 V_min = -1.0;
 V_max = 1.0;
 v0_min = -35.0;
 v0_max = 35.0;

 // Weights W are randomly distributed between W_min and
    W_max;
 W_min = -50.0; W_max = 50.0;
 // Generate the random values for V
 for (i=0; i<N1; i++) {
  for (j=0; j<N2; j++) {
   // Generate the random number between 0 and 1 for weights
     V
   V[i][j] = rand() / RAN_MAX;
   // Transform the random number to the (V_min, V_max)
     interval
   V[i][j] = (V[i][j] * (V_max - V_min)) + V_min;
  }
 }
 // Generate the random values for v0
 for (i=0; i<N2; i++) {
  // Generate the random number between 0 and 1 for weights
    v0
  v0[i] = rand() / RAN_MAX;
  // Transform the random number to the (v0_min, v0_max)
    interval
  v0[i] = (v0[i] * (v0_max - v0_min)) + v0_min;
 }
```

```c
// Generate the random values for Vi
for (i=0; i<N1i; i++) {
 for (j=0; j<N2i; j++) {
  // Generate the random number between 0 and 1 for weights
     Vi
  Vi[i][j] = rand() / RAN_MAX;
  // Transform the random number to the (V_min, V_max)
     interval
  Vi[i][j] = (Vi[i][j] * (V_max - V_min)) + V_min;
 }
}
// Generate the random values for v0i
for (i=0; i<N2i; i++) {
  // Generate the random number between 0 and 1 for weights
     v0i
  v0i[i] = rand() / RAN_MAX;
  // Transform the random number to the (v0_min, v0max)
     interval
  v0i[i] = (v0i[i] * (v0_max - v0_min)) + v0_min;
}
// INITIALIZATION
for (i=0; i<n; i++)
 x[i] = 0.0; // system states initialized to 0;

for (i=0; i<(N2+1); i++) {
 for (j=0; j<N3; j++)
  W[i][j] = 0.0; // set the states to 0;
}

for (i=0; i<(N2i+1); i++) {
 for (j=0; j<N3i; j++)
  W1i[i][j] = 0.0; // set the states to 0;
}

for (i=0; i<(Nai+1); i++) {
 for (j=0; j<N3i; j++)
  W2i[i][j] = 0.0; // set the states to 0;
}

// Joints initial position
x[0] = 0.0;
x[1] = 0.0;

/* Generate random initial values for the matrix W,
 random numbers are uniformly distributed between
 W_min and W_max. */
for (i=0; i<(N2+1); i++) {
 for (j=0; j<N3; j++) {
  W[i][j] = rand() / RAN_MAX;
```

```c
   W[i][j] = (W[i][j] * (W_max - W_min)) + W_min;
  }
 }
 // Generate matrix Wi: Wi = [W1i; W2i];
 for (i=0; i<(N2i+1); i++) {
  for (j=0; j<N3i; j++)
   Wi[i][j] = W1i[i][j];
 }
 for (i=0; i<(Nai+1); i++) {
  for (j=0; j<N3i; j++)
   Wi[N2i+1+i][j] = W2i[i][j];
 }

 // Open the files for the storage of the data
 FILE *frobot, *fweight;
 frobot = fopen("c:\\rs\\csim\\x_robot","w");
 fweight = fopen("c:\\rs\\csim\\x_weight","w");

 // main iteration loop
 k = 0;

 do {
  // finding desired path
  double amp = 1.0;
  double omega = 1.0;
  qd[0] = amp * sin(omega*t);
  // Desired position for the 1-joint
  qd[1] = amp * cos(omega*t);
  // Desired position for the 2-joint
  qdp[0] = amp * omega * cos(omega*t);
  // Desired velocity for the 1-joint
  qdp[1] = -amp * omega * sin(omega*t);
  // Desired velocity for the 2-joint
  qdpp[0] = -amp * omega * omega * sin(omega*t);
  // Desired acceleration for the 1-joint
  qdpp[1] = -amp * omega * omega * cos(omega*t);
  // Desired acceleration for the 2-joint

  // states are defined as follows: x[0] = q1; x[1] = q2;
  //   x[2] = q1dot; x[3] = q2dot

  // finding tracking errors
  e[0] = qd[0] - x[0];
  e[1] = qd[1] - x[1];
  edot[0] = qdp[0] - x[2];
  edot[1] = qdp[1] - x[3];

  // finding filtered tracking errors
  r[0] = edot[0] + Lambda[0][0]*e[0] + Lambda[0][1]*e[1];
  r[1] = edot[1] + Lambda[1][0]*e[0] + Lambda[1][1]*e[1];
```

```
// finding the norm of r
norm_r = sqrt(r[0]*r[0] + r[1]*r[1]);

// signal w
w[0] = Kv * r[0];
w[1] = Kv * r[1];

// Calculate aux2 = ViT * w + v0i;

 // find the ViT (transpose);
 double Vi_transp[N2i][N1i];
 MatrixTransp(&Vi[0][0], N1i, N2i, &Vi_transp[0][0], N2i,
 N1i);

 // find ViT * w = aux1;
 double aux1[N2i][1];
 MatrixMul(&Vi_transp[0][0], N2i, N1i, &w[0], 2, 1,
 &aux1[0][0], N2i, 1);

 // find aux2 = aux1 + v0i = ViT * w + v0i;
 double aux2[N2i][1];
 MatrixAdd(&aux1[0][0], N2i, 1, &v0i[0], N2i, 1,
 &aux2[0][0], N2i, 1);

// find the sigmoid activation function of II NN;
double sigmai1[N2i+1][1];
for (i=1; i<(N2i+1); i++)
 sigmai1[i][0] = 1.0 / (1.0 + exp(-(aux2[i-1][0])));

sigmai1[0][0] = 1.0;

// Vector function fi
if (w[0]>0) {
 fi[1] = 1.0;
 fi[2] = 1-exp(-w[0]);
}
else {
 fi[1] = 0.0;
 fi[2] = 0.0;
}
if (w[1]>0) {
 fi[3] = 1.0;
 fi[4] = 1-exp(-w[1]);
}
else {
 fi[3] = 0.0;
 fi[4] = 0.0;
}
fi[0] = 1.0;
```

```
// sigmai = [sigmai1; fi];
for (i=0; i<(N2i+1); i++)
 sigmai[i][0] = sigmai1[i][0];
for (i=0; i<(Nai+1); i++)
 sigmai[N2i+1+i][0] = fi[i];

// find NNII: NNII = WiT * sigmai;
double WiT[N3i][N2i+Nai+2];
MatrixTransp(&Wi[0][0], N2i+Nai+2, N3i, &WiT[0][0], N3i,
N2i+Nai+2);
MatrixMul( &WiT[0][0], N3i, N2i+Nai+2,
    &sigmai[0][0], Nai+N2i+2, 1,
    &NNII[0][0], N3i, 1);

// DEADZONE COMPENSATION, find signal u;
u[0] = w[0] + NNII[0][0];
u[1] = w[1] + NNII[1][0];

// Calculate aux4 = VT * u + v0;
 // find the VT (transpose);
 double V_transp[N2][N1];
 MatrixTransp(&V[0][0], N1, N2, &V_transp[0][0], N2, N1);

 // find VT * u = aux3;
 double aux3[N2][1];
 MatrixMul(&V_transp[0][0], N2, N1, &u[0], 2, 1,
     &aux3[0][0], N2, 1);

 // find aux4 = aux3 + v0 = VT * u + v0;
 double aux4[N2][1];
 MatrixAdd(&aux3[0][0], N2, 1, &v0[0], N2, 1,
     &aux4[0][0], N2, 1);

// find the sigmoid activation function of I NN;
double sigma[N2+1][1];
for (i=1; i<(N2+1); i++)
 sigma[i][0] = 1.0/(1.0 + exp(-(aux4[i-1][0])));

sigma[0][0] = 1.0;

// find sigma_prime[N2+1][N2]
double sigma_prime[N2+1][N2];
for (i=0; i<N2+1; i++) {
 for (j=0; j<N2; j++) {
  if (j == i-1)
   sigma_prime[i][j] = (sigma[j][0])*(1-(sigma[j][0]));
  else
   sigma_prime[i][j] = 0.0;
 }
}
```

```
// find NNI: NNI = WT * sigma;
double WT[N3][N2+1];
MatrixTransp(&W[0][0], N2+1, N3, &WT[0][0], N3, N2+1);
MatrixMul( &WT[0][0], N3, N2+1,
     &sigma[0][0], N2+1, 1,
     &NNI[0][0], N3, 1);

// DEADZONE in mechanical system
if (u[0] >= d_plus)
 tau[0] = m_plus*(u[0]-d_plus);
else if (u[0] <= d_minus)
 tau[0] = m_minus*(u[0]-d_minus);
else
 tau[0] = 0.0;

if (u[1] >= d_plus)
 tau[1] = m_plus*(u[1]-d_plus);
else if (u[1] <= d_minus)
 tau[1] = m_minus*(u[1]-d_minus);
else
 tau[1] = 0.0;

// Print the chosen values to the screen and file;
if (k % print_step == 0) {
 k = 0;
 // print the values to the screen;
 printf ("%f %f %f %f %f\n", t, e[0], e[1], NNII[0][0],
 NNII[1][0]);
 // print the values to the file;
 fprintf(frobot, "%f %f %f %f %f %f %f %f %f %f %f %f %f
 %f %f %f %f %f %f\n",
    t, e[0], e[1], edot[0], edot[1], qd[0], qd[1],
    NNI[0][0], NNI[1][0], NNII[0][0], NNII[1][0],
    u[0], u[1], W[0][0], W[1][3], W[5][6], Wi[0][2],
    Wi[2][5], Wi[4][7]);

 fprintf(fweight,"%f %f %f %f\n", W[0][0], W[5][6],
 Wi[0][2], Wi[4][7]);
}
k++;

// Integration for robot arm dynamics
rk4(robot_dyn, 4, &x[0], &tau[0], &x[0], t, t+time_step);

// prepare for W weights integration;
double r_transp[1][2]; // transpose of r
double tau_W[N2+1][2]; // first part of the diff. eq. for
W weights.
MatrixTransp(&r[0], 2, 1, &r_transp[0][0], 1, 2);
```

```
  // tau_W = -S*sigmap*VT*WiT*sigmai*rT;
  MatrixMul5(&sigma_prime[0][0], N2+1, N2,
        &V_transp[0][0], N2, N1,
        &WiT[0][0], N3i, N2i+Nai+2,
        &sigmai[0][0], Nai+N2i+2, 1,
        &r_transp[0][0], 1, 2,
        &tau_W[0][0], N2+1, 2);


  // Multiply result by -S
  for (i=0; i<(N2+1); i++) {
   for (j=0; j<2; j++)
    tau_W[i][j] = -tuning_S*tau_W[i][j];
  }


  // prepare for Wi weights integration;
  double tau_Wi[N2i+Nai+2][N3i];


  // tau_Wi = T*sigmai*rT*WT*sigmap*VT;
  MatrixMul5(&sigmai[0][0], Nai+N2i+2, 1,
        &r_transp[0][0], 1, 2,
        &WT[0][0], N3, N2+1,
        &sigma_prime[0][0], N2+1, N2,
        &V_transp[0][0], N2, N1,
        &tau_Wi[0][0], N2i+Nai+2, N3i);


  // Multiply result by T
  for (i=0; i<(N2i+Nai+2); i++) {
   for (j=0; j<N3i; j++)
    tau_Wi[i][j] = tuning_T*tau_Wi[i][j];
  }


  // Integration for NN weights W
  rk4(weights_W, (N2+1)*N3, &W[0][0], &tau_W[0][0], &W[0][0],
   t, t+time_step);


  // Integration for NN weights Wi
  rk4(weights_Wi, (N2i+Nai+2)*N3i, &Wi[0][0], &tau_Wi[0][0],
  &Wi[0][0],
   t, t+time_step);


  t += time_step; // increment the time;
 } while (t <= tfinal);
 fcloseall();


 return 0;
}
```

```
/*************************************************************
Function for calculating the Wdot, based on the dynamics of
the system.

INPUT VALUES: double* W, double* tau
RETURNED VALUE: double* Wdot
*************************************************************/

void weights_W(double* W, double* tau, double* Wdot) {

 extern double norm_r, k1, tuning_S;
 extern int N2_nn, N3_nn;

 int j, k;

 // Tuning law: Wdot = tau - k1*S*norm_r*W
 for (j=0; j<N2_nn+1; j++) {
  for (k=0; k<N3_nn; k++) {
   Wdot[j*N3_nn+k] = tau[j*N3_nn+k]
   - (k1*tuning_S*norm_r*(W[j*N3_nn+k]));
  }
 }
}


/*************************************************************
Function for calculating the Widot, based on the dynamics of
the system.

INPUT VALUES: double* Wi, double* tau
RETURNED VALUE: double* Widot
*************************************************************/

void weights_Wi(double* Wi, double* tau, double* Widot) {

 extern double norm_r, k1, k2, tuning_T;
 extern int N2i_nn, Nai_nn, N3i_nn;

 int j, k;
 double sum, norm_Wi;

 // Tuning law: Widot = tau - k1*T*norm_r*Wi
 //    - k2*T*norm(Wi)*Wi

 // Finding the norm(Wi);
 sum = 0.0;
 for (j=0; j<N2i_nn+Nai_nn+2; j++) {
  for (k=0; k<N3i_nn; k++) {
   sum = sum + (Wi[j*N3i_nn+k])*(Wi[j*N3i_nn+k]);
  }
 }
```

```
 norm_Wi = sqrt(sum);

 for (j=0; j<N2i_nn+Nai_nn+2; j++) {
  for (k=0; k<N3i_nn; k++) {
   Widot[j*N3i_nn+k] = tau[j*N3i_nn+k]
     - (k1*tuning_T*norm_r*(Wi[j*N3i_nn+k]))
     - (k2*tuning_T*norm_r*norm_Wi*(Wi[j*N3i_nn+k]));
  }
 }
}

/************************************************************
Function for calculating the xdot, based on the dynamics of
the system.

INPUT VALUES: double* x, double* tau
RETURNED VALUE: double* xdot
************************************************************/

void robot_dyn(double* x, double* tau, double* xdot) {

 // Parameters of the robot dynamics
 double m11, m12, m22, v1, v2, a1, a2, g1, g2, det;
 double m1=1.8, m2=1.3, le1=1.0, le2=1.0, g=9.8;

 double x0, x1, x2, x3, tau0, tau1;

 x0 = *x;
 x1 = *(x+1);
 x2 = *(x+2);
 x3 = *(x+3);

 tau0 = *tau;
 tau1 = *(tau+1);

 // Robot Dynamics
 m11 = (m1+m2)*le1*le1 + m2*le2*le2 + 2.*m2*le1*le2*cos(x1);
 m22 = m2*le2*le2;
 m12 = m22 + m2*le1*le2*cos(x1);

 v1 = -m2*le1*le2*(2.*x2*x3 + x3*x3)*sin(x1);
 v2 = m2*le1*le2*x2*x2*sin(x1);

 g1 = (m1+m2)*g*le1*cos(x0) + m2*g*le2*cos(x0 + x1);
 g2 = m2*g*le2*cos(x0 + x1);

 det = m11*m22 - m12*m12;

 // gravity term included;
```

```
  a1 = tau0 - v1 - g1;
  a2 = tau1 - v2 - g2;

  // State-space model of the robot
  *xdot = x2;
  *(xdot+1) = x3;
  *(xdot+2) = ( m22*a1 - m12*a2 )/det;
  *(xdot+3) = ( m11*a2 - m12*a1 )/det;
}

// Runge Kutta 4. order
void rk4 (void (*rastko)(double* x, double* tau,
double* xdot),
    int states_num,
    double* x_in,
    double* u,
    double* x_out,
    double t0,
    double tf) {

  // Runge Kutta for integrating the differential equations
  // Usage:
  //   1. file which contains the dynamics of the system,
  //   2. number of states for integration;
  //   3. x_in - input states
  //   4. u - additional input
  //   5. x_out - output states after the step of integration
  //   6. t0 - starting time
  //   7. tf - final time

 int i;
 double h = tf-t0;
 double h2 = h/2.0;
 double h3 = h/3.0;
 double h6 = h/6.0;
 double *x_mid1, *x_mid2, *x_end, *xdot_init, *xdot_mid1;
 double *xdot_mid2, *xdot_end;

 x_mid1 = (double *) calloc(states_num, sizeof(double));
 x_mid2 = (double *) calloc(states_num, sizeof(double));
 x_end = (double *) calloc(states_num, sizeof(double));
 xdot_init = (double *) calloc(states_num, sizeof(double));
 xdot_mid1 = (double *) calloc(states_num, sizeof(double));
 xdot_mid2 = (double *) calloc(states_num, sizeof(double));
 xdot_end = (double *) calloc(states_num, sizeof(double));

 (*rastko)(x_in, u, xdot_init); // get xdot at initial x

 for (i=0; i<states_num; i++) x_mid1[i] = x_in[i]
```

```
 + h2*xdot_init[i];
 (*rastko)(x_mid1, u, xdot_mid1); // get xdot at first
 guess midpt

 for (i=0; i<states_num; i++) x_mid2[i] = x_in[i]
 + h2*xdot_mid1[i];
 (*rastko)(x_mid2, u, xdot_mid2); // get xdot at 2nd guess
 midpt

 for (i=0; i<states_num; i++) x_end[i] = x_in[i]
 + h*xdot_mid2[i];
 (*rastko)(x_end, u, xdot_end);
 // get xdot at estimated x(t+dt)

 for (i=0; i<states_num; i++)
  x_out[i] = x_in[i] + h6*(xdot_init[i]+xdot_end[i])
  + h3*(xdot_mid1[i]+xdot_mid2[i]);

 free(x_mid1); free(x_mid2); free(x_end); free(xdot_init);
 free(xdot_mid1); free(xdot_mid2); free(xdot_end);
}

/*************************************************************
Functions for matrix manipulation: transpose, multiplication,
and summation.

Author: Rastko R. Selmic

July 19, 1998.
The University of Texas at Arlington
*************************************************************/

/*************************************************************
Sum of 2 matrices

This function adds the matrix A1(m1, n1) with matrix
A2(m1, n1). The result is the matrix R(m1, n1).
*************************************************************/

void MatrixAdd (double *A1, int m1, int n1,
       double *A2, int m2, int n2,
       double *R, int p, int q) {

 int j, k;

 for (j=0; j<m1; j++) {
  for (k=0; k<n1; k++) {
   *(R + j*n1+k) = *(A1 + j*n1+k) + *(A2 + j*n1+k);
  }
 }
}
```

```
/***********************************************************
Multiplication of the matrices

This function multiplies the matrix A1(m1, n1) with matrix
A2(m2, n2). The result is the matrix R(m1, n2).
***********************************************************/

void MatrixMul (double *A1, int m1, int n1,
        double *A2, int m2, int n2,
        double *R, int p, int q) {

 int j, k, l;
 double sum;

 for (j=0; j<m1; j++) {
  for (k=0; k<n2; k++) {
   sum = 0;
   for (l=0; l<n1; l++) {
    sum = sum + (*(A1 + j*n1+l)) * (*(A2 + l*n2+k));
   }
   *(R + j*q+k) = sum;
  }
 }
}


/***********************************************************
Transpose of matrix

This function transposes matrix A1(m,n). The result is the
matrix R(n,m).
***********************************************************/

void MatrixTransp (double *A1, int m, int n,
         double *R, int p, int q) {

 int j, k;

 for (j=0; j<m; j++) {
  for (k=0; k<n; k++) {
   *(R + k*m+j) = *(A1 + j*n+k);
  }
 }
}


/***********************************************************
Multiplication of the three matrices

This function multiplies the matrices A1(m1, n1), A2(m2, n2),
and A3(m3, n3). The result is the matrix R(m1, n3).
***********************************************************/
```

```
void MatrixMul3 (double *A1, int m1, int n1,
        double *A2, int m2, int n2,
        double *A3, int m3, int n3,
        double *R, int p, int q) {

  int p_temp, q_temp;
  double* Temp;

  p_temp = m1;
  q_temp = n2;

  Temp = (double *) calloc(p_temp*q_temp, sizeof(double));

  MatrixMul (A1, m1, n1,
     A2, m2, n2,
     Temp, p_temp, q_temp);

  MatrixMul (Temp, p_temp, q_temp,
      A3, m3, n3,
      R, p, q);
  free (Temp);
}


/*************************************************************
Multiplication of four matrices

This function multiplies the matrices A1(m1, n1), A2(m2, n2),
A3(m3, n3), A4(m4, n4). The result is the matrix R(m1, n4).
*************************************************************/

void MatrixMul4 (double *A1, int m1, int n1,
        double *A2, int m2, int n2,
        double *A3, int m3, int n3,
        double *A4, int m4, int n4,
        double *R, int p, int q) {

  int p_temp, q_temp;
  double* Temp;

  p_temp = m1;
  q_temp = n3;

  Temp = (double *) calloc(p_temp*q_temp, sizeof(double));

  MatrixMul3 (A1, m1, n1,
      A2, m2, n2,
      A3, m3, n3,
      Temp, p_temp, q_temp);
```

```
 MatrixMul (Temp, p_temp, q_temp,
      A4, m4, n4,
      R, p, q);
 free (Temp);
}


/************************************************************
This function multiplies the matrices A1(m1, n1), A2(m2, n2),
A3(m3, n3), A4(m4, n4), A5(m5, n5). The result is the matrix
R(m1, n4).
************************************************************/

void MatrixMul5 (double *A1, int m1, int n1,
        double *A2, int m2, int n2,
        double *A3, int m3, int n3,
        double *A4, int m4, int n4,
        double *A5, int m5, int n5,
        double *R, int p, int q) {

 int p_temp, q_temp;
 double* Temp;

 p_temp = m1;
 q_temp = n4;

 Temp = (double *) calloc(p_temp*q_temp, sizeof(double));

 MatrixMul4 (A1, m1, n1,
      A2, m2, n2,
      A3, m3, n3,
      A4, m4, n4,
      Temp, p_temp, q_temp);

 MatrixMul (Temp, p_temp, q_temp,
      A5, m5, n5,
      R, p, q);
 free (Temp);
}
```

Appendix C

C Code for Discrete-Time Neural Network Backlash Controller

This appendix contains the C code for the discrete-time NN backlash controller simulation in Chapter 5. There are two files: a header file and a program source code file. For more information, see Campos Portillo (2000).

Header file

```
/*
 /* backsim.h

 Header file for discrete-time backlash nonlinear system
 simulation

 Javier Campos, November 01, 2000

*/

#ifndef _BACKSIM
#define _BACKSIM

#include <math.h>

/* Simulation Constant Definitions */
#define pi   3.141592653589793

/* Dynamics Parameters */

/* Controller Parameters */
#define Kv 0.285
#define Kb -2.0
#define lamda1 0.25

 /* PARAMETERS OF THE NN */
#define N1 7 /*  Input layer */
```

```
#define N2 10 /*  Hidden layer */
#define N3 1 /*  Output layer */

#define alpha 0.1    //0.1
#define gamma 0.2    //0.01

// #define WMax 100000.0  // Use for saturate weights if
   needed

 /* desired input */
#define amplitude 1.0
#define period 4.0*pi // seconds

 /* backlash nonlinearity parameters */
#define m 0.5  // slope
#define d_plus 0.2 // 0.2 default
#define d_minus 0.2 // 0.2 default

/* Simulation run-time definitions */
double T_step = 0.001; // Time step (secs)
double T = 15.0; // Simulation run time (secs)
int print_step = 100;

/* Initial conditions */
#define u_init 0.0 // u
#define x1_init 0.0 //
#define x2_init 0.0 //
#define tau_init 0.0

/* Function Prototypes */
void rk4(void (*rk)(double x[], double x_d[], double xdot[]),
  int order, double x_in[], double u[], double x_out[],
  double t0, double tf);

#endif /* ifndef _BACKSIM */
```

Source file

```
/* backsim.c

 Nonlinear backlash Simulation source file.

 Javier Campos,  November 01, 2000

*/

/*  To avoid confusion, all variables begin their indices at
    1, i.e., x[1]... */
```

```c
#include "backsim.h"
#include <stdio.h>
#include <math.h>
#include <stdlib.h>
#include <malloc.h>

double NN=0.0;

/**************   Main Simulation Procedure   ***************/
void main()
{
 int step = 0;
 int k = print_step;
 FILE *data_x1;
 double t; // time variables
 double norm; // 2-norm for tuning law
 double x1,x2; // state variables for time k
 double x1_kp1,x2_kp1;  // state variables for time k+1
 double x_desired1, x_desired2; // desired tracking state for
 time t
 double x_desired1_kp1, x_desired2_kp1; // desired tracking
 state for time t+k
 double x_desired1_kp2, x_desired2_kp2; // desired tracking
 state for time t+2k
 double f; // nonlinear function estimate for time k

 double w;

 double e1,e2; // tracking errors for time k
 double e1_kp1,e2_kp1; // tracking errors for time k+1

 double r;//filtered error for time k
 double r_kp1;//filtered error for time k+1


 int i,j;
 double tau_des,tau,u; // actuator input,backlash input for
 time k
 double tau_kp1,u_kp1; // actuator input,backlash input for
 time k+1

 double auxnn[N2]; // temp array variable

 double weight[N3*(N2+1)];         //states
 double V1[N1][N2],v0[N2];   //NN
 double sigma[N2+1];     //hidden layer neurons
 double V_min, V_max, v0_min, v0_max, w_min, w_max; //init
 double RAN_MAX = 32676.0;
 double X1[N1+1];     //NN input layers
```

```
double sq, ident;
double L, L_km1, a;      // discrete filter parameters

a = 0.45;

t = 0.0; // initial time

w=0.5; // rad/second   //desired input frequency

// Open simulation data file
data_x1=fopen("data1.txt","w");

   // Initial Conditions of the system

x1 = x1_init;
   x2 = x2_init;

NN=0.0;

u = u_init;
tau = 0.0;

L_km1 = 0.0;

/*  random numbers are uniformly distributed between v_min
    and v_max */
V_min = -0.1;
V_max = 0.1;
v0_min = -100.0;
v0_max = 100.0;


 /*  Generate the random values for V and v0*/

for(i=0; i<N1; i++)
{
 for(j=0; j<N2; j++)
 {
  /* generate random number between 0 and 1 for input
     weights V */
  V1[i][j] = rand()/RAN_MAX;
  /* transform random number to the V_min and V_max
     interval */
  V1[i][j] = (V1[i][j]*(V_max - V_min))+V_min;
 }
}
for(i=0; i<N2; i++)
{
 /* generate random number between 0 and 1 for thresholds */
```

```
 v0[i] = rand()/RAN_MAX;
 /* transform random number to the v0_min and v0_max
    interval */
 v0[i] = (v0[i]*(v0_max - v0_min))+v0_min;
}

/* Initialize activation function matrices for NN */
for(i=0;i<(N2+1);i++)
{
 /* initialize sigmoid */
 sigma[i] = 0;
}

/* Initialize auxiliary variable */
for(i=0; i<N2; i++)
{
 auxnn[i] = 0.0;
}

/* Initialize all weight states to zero */
for(i=0; i< (N3*(N2+1)); i++)
 weight[i]=0.0; /* NN states */

  /* Generate random initial values for the matrix weights
 random numbers are uniformly distributed between w_min and
 w_max. */
 w_min = -100.0;
 w_max = 100.0;

for(j=0; j<(N2+1); j++)
 weight[j] = (rand()/RAN_MAX)*(w_max - w_min)+w_min;

do{

 //***  State-space equations ***//
 // Calculate the states for time k+1
 x1_kp1 = x2;
 x2_kp1 = -0.1875*x1/(1+x2*x2)+x2+tau;

 // Create the desired trajectory for time k
 x_desired1 = amplitude*cos(w*(t-T_step));
 x_desired2 = amplitude*cos(w*t);

 // Create the desired trajectory for time k+1
 x_desired1_kp1 = amplitude*cos(w*t);
 x_desired2_kp1 = amplitude*cos(w*(t+T_step));

 // Create the desired trajectory for time k+2
 x_desired1_kp2 = amplitude*cos(w*(t+T_step));
```

```
x_desired2_kp2 = amplitude*cos(w*(t+2.0*T_step));

// Calculate tracking errors for time k
e1=x1-x_desired1;
e2=x2-x_desired2;

// Calculate tracking errors for time k+1
e1_kp1=x1_kp1-x_desired1_kp1;
e2_kp1=x2_kp1-x_desired2_kp1;

// Calculate filtered error for time k
r=e2+lamda1*e1;

// Calculate filtered error for time k+1
r_kp1=e2_kp1+lamda1*e1_kp1;

// Calculate nonlinear function for time t
f=-0.1875*x1/(1+x2*x2)+x2;

// Calculate tau_des for time t
tau_des=Kv*r - f + x_desired2_kp1 - lamda1*e1;

// Calculate discrete time filter output
L = a*L_km1+a*tau_des;

/* NN Input Vector */
X1[0] = 1.0;
X1[1] = x1;
X1[2] = x2;
X1[3] = r;
X1[4] = x_desired1;
X1[5] = x_desired2;
X1[6] = tau;
X1[7] = tau_des-tau;

/* Computation of sigmoid function for NN */
for(i=0; i<N2; i++)
{
 auxnn[i] = v0[i]*X1[0];
 for(j=0;j<N1;j++)
  auxnn[i] += V1[j][i]*X1[j+1];
}
for(i=1; i<(N2+1); i++)
{
 sigma[i] = 1.0/(1.0+exp(-auxnn[i-1]));
}
sigma[0] = 1.0;

// saturate weights if necessary
```

```
/*for(j=0;j<(N2+1);j++)
{
 if (weight[j] > WMax)
  weight[j] = WMax;
 else if (weight[j] < -WMax)
  weight[j] = -WMax;
}*/

/* NN  output */
NN = 0.0;
for(j=0;j<(N2+1);j++)
  NN = NN + weight[j]*sigma[j];
//NN = temp;


// Calculate u for time k+1
u_kp1= -Kb*(tau_des-tau)+L+NN;

// This is for backlash without compensation. Comment this
   out for backlash compensation
//u_kp1 = tau_des;

// Apply Backlash

if (((u_kp1 > 0.0) && (u=m*tau-m*d_plus))
|| ((u_kp1 < 0.0) && (u=m*tau-m*d_minus)))
 tau_kp1=m*u_kp1;
else
 tau_kp1=0.0;

// No backlash. ** Comment this out for backlash **
//tau_kp1 = tau_des;

// Print current values
if(k%print_step==0){
 k=0;
 printf("t=%f %f %f %f %f %f %f %f %g\n", t, x1, x2, e1,
 e2, r, u, tau, NN);
 fprintf(data_x1,"%f %g %g %g %g %g %g %g %g\n",
  t, x1, x2,e1, e2, r, u,tau, NN);
} //close if

k++;

/* NN tuning rules */
/* Dif. eq. for W */

norm = 0.0;
for(i=0; i<N2;i++)
```

```c
  for(j=0; j<N2; j++)
  {
   // create identity matrix of size (N2+1)x(N2+1)
   ident = 0.0;
   if (i==j)
    ident = 1.0;
   // create I-alpha*sigma*sigma^T matrix
   sq = ident - alpha*sigma[i+1]*sigma[j+1];
   // find the 2-norm of the I-alpha*sigma*sigma^T matrix
   norm+=sq*sq;
  }

 norm=sqrt(norm);

 for(j=0;j<(N2+1);j++)
  weight[j] = weight[j]
  + alpha*sigma[j]*(r_kp1+tau_kp1)-gamma*weight[j]*norm;

 // prepare for next iteration
 x1=x1_kp1;
 x2=x2_kp1;
 u=u_kp1;
 tau=tau_kp1;

 L_km1 = L;

 step++;
 t += T_step;
}while(t <= T);
fcloseall();

} // close main function

void rk4(void (*rk)(double x[], double x_d[], double xdot[]),
  int order, double x_in[], double u[], double x_out[],
  double t0, double tf)
  // Runge Kutta for integrating the differential equations
  // Usage:
  //  1. file which contains the dynamics of the system,
  //  2. number of the states
  //  3. x_in - input states
  //  4. u - additional input
  //  5. x_out - output states after the step of integration
  //  6. t0 - starting time
  //  7. tf - final time
{
 int i;
    double h = tf-t0;
    double h2 = h/2.0;
```

```
    double h3 = h/3.0;
    double h6 = h/6.0;
    double *x_mid1, *x_mid2, *x_end, *xdot_init, *xdot_mid1;
    double *xdot_mid2, *xdot_end;

    x_mid1=calloc(order+1,sizeof(double));
    x_mid2=calloc(order+1,sizeof(double));
    x_end=calloc(order+1,sizeof(double));
    xdot_init=calloc(order+1,sizeof(double));
    xdot_mid1=calloc(order+1,sizeof(double));
    xdot_end=calloc(order+1,sizeof(double));
    xdot_mid2=calloc(order+1,sizeof(double));

    (*rk)(x_in,u,xdot_init); /* get xdot at initial x */
    for (i=1;i<=order;i++)
  x_mid1[i] = x_in[i] + h2*xdot_init[i];
    (*rk)(x_mid1,u,xdot_mid1);/* get xdot at first guess
    midpt*/
    for (i=1;i<=order;i++)
  x_mid2[i] = x_in[i] + h2*xdot_mid1[i];
    (*rk)(x_mid2,u,xdot_mid2);/* get xdot at 2nd guess
    midpt */
    for (i=1;i<=order;i++)
  x_end[i] = x_in[i] + h*xdot_mid2[i];
    (*rk)(x_end,u,xdot_end);/* get xdot at estimated
    x(t+dt) */
    for (i=1;i<=order;i++)
      x_out[i] = x_in[i] + h6*(xdot_init[i]+xdot_end[i])
  + h3*(xdot_mid1[i]+xdot_mid2[i]);
    free(x_mid1);free(x_mid2);free(x_end);free(xdot_init);
    free(xdot_mid1);free(xdot_mid2);free(xdot_end);
} // close rk4
```

Appendix D

Versatile Real-Time Executive Code for Implementation of Neural Network Backstepping Controller on ATB1000 Tank Gun Barrel

This appendix contains the VRTX code used in Chapter 9 for implementing the NN backstepping controller on the ATB1000 tank gun barrel testbed. The code allows implementation on the PC–PC controller developed at UTA (Ikenaga (2000)) and described in Chapter 9.

Neural network PC–PC code

In the following, the NN flexible system control algorithm is given for implementation on the PC–PC controller. This algorithm was written under the PC–PC philosophy described in Chapter 9. The control algorithm routines running on the PC–PC target computer are written in a modular fashion using standard C. This scheme simplifies advanced control algorithm prototyping. Each control system module or block written for this architecture includes six different operation modes. This gives the user full control over the module operation at any time. Also, it uses PC–PC standard structure types and function descriptions. Included here is only the code for the NN controller. Also required is interfacing code and initialization code for the VRTX controller. More details are available from the University of Texas at Arlington (flewis@uta.edu).

```
/* Neural network default parameters*/
const nl_nnet_parm_t nl_nnet_parm_def = {
 /* Parameters for NN */
 1.0, /* Rigid Control position gain */
 1.0, /* Rigid Control velocity gain */
 2.0, /* NN e-mod rate (Kappa) */
 0.000001, /* NN learning rate for W (F) */
 20.0, /* NN learning rate for V (G) */
 100.0, /* LamdaMax */
 1.0e15, /* input weights saturation */
 1.0e9, /* output weights saturation */
};

/* Neural network algorithm */
int nl_nnet(int mode, int reset, int size, double dt,
 const nl_nnet_parm_t *parm,
  nl_nnet_state_t *state, const nl_nnet_in_t *in,
```

```
  nl_nnet_out_t *out)
{
 int i,j,k;
 double lamda, KappaSgnFltre, Ts_F_Fltre, Ts_G_Fltre,
 Alpha=1.0;
 double H[N2]; /* Input to hidden layer */
 double Sigma[N2];  /* Hidden layer output */
 double SigmaPrime[N2]; /* Derivatives of hidden layers */
 double fltre; /* Filtered error */

 if (reset)
 mode = SB_MODE_RESET;

 for (i = 0; i < size; i++) {
 switch (mode) {
 case SB_MODE_ON:
 case SB_MODE_FREEZE:
 case SB_MODE_BYPASS:
     {
 if(parm[i].kv)
 lamda = parm[i].kp/parm[i].kv;
 else
 lamda = 0.0;
 if(!parm[i].kv || lamda > parm[i].lamdaMax)
 lamda=parm[i].lamdaMax;


 /* Calculate Filtered error */
 fltre = in[N1*i+1] + lamda*in[N1*i];

 /* Calculate H, Sigma, and SigmaPrime */
 for(j=0;j<N2;j++) {
  H[j]=state[i].V[j];
  for(k=1;k<=N1;k++) {
 H[j]+=state[i].V[k*N2+j]*in[N1*i+k-1];
  }
  Sigma[j]=1.0/(1.0 + exp(-(j+1.0)*Alpha*H[j]));
  SigmaPrime[j]=(j+1.0)*Alpha*Sigma[j]*(1.0 - Sigma[j]);
 }
 /* Update rules for weight matrices V and W */

 /* Consider sign of filtered error */
 KappaSgnFltre = (fltre >=0.0)
 ? parm[i].kappa:-parm[i].kappa;

 /* Optimize operations */
 Ts_G_Fltre = dt*parm[i].G*fltre*0.5;
 Ts_F_Fltre = dt*parm[i].F*fltre*0.5;
  }
```

```
 break;
}
switch (mode) { /* output result */
case SB_MODE_ON:
case SB_MODE_FREEZE:
{
 /* Calculate Neural Network output */
out[i]=state[i].W[0];
for(j=1;j<=N2;j++) {
out[i]+=state[i].W[j]*Sigma[j-1];
}
}
break;

case SB_MODE_BYPASS:
case SB_MODE_SHORT:
{
out[i] = in[i];
}
break;

case SB_MODE_RESET:
case SB_MODE_OFF:
case SB_MODE_DISABLE:
default:
{
out[i] = 0.0;
}
break;
}
switch (mode) { /* update state */
case SB_MODE_ON:
case SB_MODE_OFF:
case SB_MODE_BYPASS:
{
/* Update input layer weights V */
for (j=0;j<N2;j++) {
  state[i].V[j]+=state[i].DeltaV[j];
       state[i].DeltaV[j]=Ts_G_Fltre*(state[i].W[j+1]*
        SigmaPrime[j]-KappaSgnFltre*state[i].V[j]);
  state[i].V[j]+=state[i].DeltaV[j];
for(k=1;k<=N1;k++) {
 state[i].V[k*N2+j]+=state[i].DeltaV[k*N2+j];
       state[i].DeltaV[k*N2+j]=Ts_G_Fltre*(in[N1*I
        +k-1]*state[i].W[j+1]*SigmaPrime[j]
  - KappaSgnFltre*state[i].V[k*N2+j]);
       state[i].V[k*N2+j]+=state[i].DeltaV[k*N2+j];
}
}
```

```
for(j=0;j<((N1+1)*N2);j++) {
        /* Saturate input weights */
state[i].V[j] = satw(state[i].V[j], parm[i].vMax);
}
/* Update output layer weights W */
state[i].W[0]+=state[i].DeltaW[0];
state[i].DeltaW[0]=Ts_F_Fltre*(1.0-KappaSgnFltre*
          state[i].W[0]);
state[i].W[0]+=state[i].DeltaW[0];

for(k=0;k<N2;k++) {
state[i].W[k+1]+=state[i].DeltaW[k+1];
state[i].DeltaW[k+1]=Ts_F_Fltre*(Sigma[k]-
        SigmaPrime[k]*H[k]
 -KappaSgnFltre*state[i].W[k+1]);
state[i].W[k+1]+=state[i].DeltaW[k+1];
}
for(j=0;j<(N2+1);j++) {
/* Saturate input weights */
state[i].W[j]=satw(state[i].W[j], parm[i].wMax);
}
}
break;

case SB_MODE_RESET:
{
for(j=0;j<(N2+1);j++) {
state[i].W[j] = 0.0;
state[i].DeltaW[j] = 0.0;
}
for(j=0;j<((N1+1)*N2);j++) {
state[i].V[j] = 0.0;
state[i].DeltaV[j] = 0.0;
}
}
break;
}
}
return 0; /* no error */
}
```

References

J. S. Albus, A new approach to manipulator control: The Cerebellar Model Articulation Cortex Controller (CMAC), *Trans. ASME J. Dynam. Systems Meas. Control*, 93 (1975), pp. 220–227.

G. Alevisakis and D. E. Seborg, An extension of Smith predictor method to multivariable linear systems containing time delays, *Internat. J. Control*, 3 (1973), pp. 541–551.

A. Alleyne and J. K. Hedrick, Nonlinear control of a quarter car active suspension, in *Proceeding of the American Control Conference*, Chicago, IEEE, Piscataway, NJ, 1982, pp. 21–25.

A. Alleyne and J. K. Hedrick, Nonlinear adaptive control of active suspensions, *IEEE Trans. Control Systems Tech.*, 3 (1995), pp. 94–101.

A. Alleyne, R. Liu, and H. Wright, On the limitations of force tracking control for hydraulic active suspensions, in *Proceedings of the American Control Conference*, Philadelphia, IEEE, Piscataway, NJ, 1998, pp. 43–47.

R. J. Anderson and M. W. Spong, Bilateral control of teleoperators with time delay, *IEEE Trans. Automat. Control*, AC-34 (1989), pp. 494–501.

B. Armstrong-Hélouvry, P. Dupont, and C. Canudas de Wit, A survey of models, analysis tools and compensation methods for the control of machines with friction, *Automatica*, 30 (1994), pp. 1083–1138.

A. R. Barron, Universal approximation bounds for superposition of sigmoidal function, *IEEE Trans. Inform. Theory*, 39 (1993), pp. 930–945.

B. W. Bequette, Nonlinear control of chemical processes: A review, *Indust. Engrg. Chem. Res.*, 30 (1991), pp. 1391–1413.

L. A. Bernotas, P. E. Crago, and H. J. Chizeck, Adaptive control of electrically stimulated muscle, *IEEE Trans. Biomed. Engrg.*, BME-34 (1987), pp. 140–147.

N. K. Bose and P. Liang, *Neural Network Fundamentals*, McGraw–Hill, New York, 1996.

F. T. Buzan and T. B. Sheridan, A model-based predictive operator aid for telemanipulators with time delay, in *Proceedings of the International Conference on Systems, Man, and Cybernetics*, 1989, pp. 138–143.

C. I. Byrnes and W. Lin, Losslessness, feedback equivalence, and the global stabilization of discrete-time nonlinear systems, *IEEE Trans. Automat. Control*, 39 (1994), pp. 83–98.

B. CAI AND D. KONIK, Intelligent vehicle active suspension control using fuzzy logic, in *12th IFAC Triennial World Congress*, Sydney, Pergamon Press, Oxford, 1993, pp. 51–56.

J. CAMPOS PORTILLO, *Intelligent Control of Complex Mechanical Systems*, Ph.D. thesis, University of Texas at Arlington, Arlington, TX, 2000.

C. CANUDAS DE WIT, P. NOËL, A. AUBIN, AND B. BROGLIATO, Adaptive friction compensation in robot manipulators: Low velocities, *Internat. J. Robotics Res.*, 10 (1991).

C. CANUDAS DE WIT, H. OLSSON, K. J. ÅSTRÖM, AND P. LISCHINSKY, A new model for control of systems with friction, *IEEE Trans. Automat. Control*, 40 (1995), pp. 419–425.

F. C. CHEN AND H. K. KHALIL, Adaptive control of nonlinear systems using neural networks, *Internat. J. Control*, 55 (1992), pp. 1299–1317.

Y. CHENG, T. W. KARJALA, AND D. M. HIMMELBLAU, Identification of nonlinear dynamic processes with unknown and variable dead time using an internal recurrent neural network, *Indust. Engrg. Chem. Res.*, 34 (1995), pp. 1735–1742.

S. COMMURI, *A Framework for Intelligent Control of Nonlinear Systems*, Ph.D. thesis, University of Texas at Arlington, Arlington, TX, 1996.

M. J. CORLESS AND G. LEITMANN, Continuous state feedback guaranteeing uniform ultimate boundedness for uncertain dynamic systems, *IEEE Trans. Automat. Control*, 26 (1982), pp. 850–861.

J. J. CRAIG, *Adaptive Control of Robot Manipulators*, Addison–Wesley, Reading, MA, 1988.

G. CYBENKO, Approximation by superpositions of a sigmoidal function, *Math. Control Signals Systems*, 2 (1989), pp. 303–314.

H. DEMUTH AND M. BEALE, *Neural Network Toolbox*, The MathWorks, Natick, MA, 1992.

C. DESOER AND S. M. SHAHRUZ, Stability of dithered nonlinear systems with backlash or hysteresis, *Internat. J. Control*, 43 (1986), pp. 1045–1060.

H. DU AND S. S. NAIR, Low velocity friction compensation, *IEEE Control Systems Magazine*, 18 (1998), pp. 61–69.

M. K. EILER AND F. B. HOOGTERP, Analysis of active suspension controllers, in *Computer Simulation Conference*, San Diego, Society for Computer Simulation International, San Diego, 1994, pp. 274–279.

G. H. ENGELMAN AND G. RIZZONI, Including the force generation process in active suspension control formulation, in *Proceedings of the American Control Conference*, San Francisco, 1993.

D. ENNS, D. BUGAJSKI, R. HENDRICK, AND G. STEIN, Dynamic inversion: An evolving methodology for flight control design, *Internat. J. Control*, 59 (1994), pp. 71–91.

T. FUKAO, A. YAMAWAKI, AND N. ADACHI, Active control of partially known systems using backstepping: Application to H_∞ design of active suspensions, in *Proceedings of the 37th Conference on Decision and Control*, IEEE Computer Society Press, Los Alamitos, CA, 1998, pp. 481–486.

T. P. GOODMAN, How to calculate dynamic effects of backlash, *Machine Design*, 1963, pp. 150–157.

M. GRUNDELIUS AND D. ANGELLI, Adaptive control of systems with backlash acting on the input, in *Proceedings of the 35th Conference on Decision and Control*, IEEE Computer Society Press, Los Alamitos, CA, 1996, pp. 4689–4694.

V. GULLAPALLI, J. A. FRANKLIN, AND H. BENBRAHUIM, Acquiring robot skills via reinforcement learning, *IEEE Control Systems Magazine*, 4 (1994), pp. 13–24.

W. M. HADDAD, J. L. FAUSZ, AND C. ABDALLAH, Optimal discrete-time control for nonlinear cascade systems, *J. Franklin Inst. B*, 335 (1998), pp. 827–839.

B. HANNAFORD, A design framework for teleoperators with kinesthetic feedback, *IEEE Trans. Robotics Automat.*, 5 (1989), pp. 426–434.

B. HANNAFORD AND W. S. KIM, Force reflection, shared control, and time delay in telemanipulation, in *Proceedings of the International Conference on Systems, Man, and Cybernetics*, 1989, pp. 133–137.

S. HAYKIN, *Neural Networks*, IEEE Press and Macmillan, New York, 1994.

M. A. HENSON AND D. E. SEBORG, Time delay compensation for nonlinear processes, *Indust. Engrg. Chem. Res.*, 33 (1994), pp. 1493–1500.

K. HORNIK, M. STINCHCOMBE, AND H. WHITE, Multilayer feedforward networks are universal approximators, *Neural Networks*, 2 (1985), pp. 359–366.

B. IGELNIK AND Y. H. PAO, stochastic choice of basis functions in adaptive function approximation and the functional-link net, *IEEE Trans. Neural Networks*, 6 (1995), pp. 1320–1329.

S. IKENAGA, *Development of a Real Time Digital Controller: Application to Active Suspension Control of Ground Vehicles*, Ph.D thesis, Department of Electrical Engineering, University of Texas at Arlington, Arlington, TX, 2000.

S. IKENAGA, F. L. LEWIS, L. DAVIS, J. CAMPOS, M. EVANS, AND S. SCULLY, Active suspension control using a novel strut and active filtered feedback: Design and implementation, in *Proceedings of the Conference on Control Applications*, Kona, HI, 1999, pp. 1502–1508.

P. A. IOANNOU AND A. DATTA, Robust adaptive control: A unified approach, *Proc. IEEE*, 790 (1991), pp. 1736–1768.

P. A. IOANNOU AND P. V. KOKOTOVIĆ, Instability analysis and improvement of robustness of adaptive control, *Automatica*, 3 (1984), pp. 583–594.

S. JAGANNATHAN, Robust backstepping control of robotic systems using neural networks, in *Proceedings of the 37th IEEE Conference on Decision and Control*, IEEE Computer Society Press, Los Alamitos, CA, 1998, pp. 943–948.

S. JAGANNATHAN AND F. L. LEWIS, Discrete-time control of a class of nonlinear dynamical systems, *Internat. J. Intell. Control Systems*, 1 (1996), pp. 297–326.

I. KANELLAKOPOULOS, P. V. KOKOTOVIĆ, AND A. S. MORSE, Systematic design of adaptive controllers for feedback linearizable systems, *IEEE Trans. Automat. Control*, 36 (1991), pp. 1241–1253.

A. KARAKASOGLU, S. I. SUDHARSANAN, AND M. K. SUNDARESHAN, Identification and decentralized adaptive control using dynamical neural networks with application to robotic manipulators, *IEEE Trans. Neural Networks*, 4 (1993), pp. 919–930.

D. KARNOPP, Active damping in road vehicle suspension systems, *Vehicle System Dynam.*, 12 (1983), pp. 291–316.

D. KARNOPP, Passive and active control of road vehicle heave and pitch motion, in *IFAC 10th Triennial World Congress*, 1987, pp. 183–188.

D. KARNOPP AND G. HEESS, Electronically controllable vehicle suspensions, *Vehicle System Dynam.*, 20 (1991), pp. 207–217.

B. S. KIM AND A. J. CALISE, Nonlinear flight control using neural networks, *J. Guidance Control Dynam.*, 20 (1997).

J.-H. KIM, J.-H. PARK, S.-W. LEE, AND E. K. P. CHONG, Fuzzy precompensation of PD controllers for systems with deadzones, *J. Intel. Fuzzy Systems*, 1 (1993), pp. 125–133.

J.-H. KIM, J.-H. PARK, S.-W. LEE, AND E. K. P. CHONG, A two-layered fuzzy logic controller for systems with deadzones, *IEEE Trans. Indust. Electron.*, 41 (1994), pp. 155–162.

Y. H. KIM AND F. L. LEWIS, Direct-reinforcement-adaptive-learning network control for nonlinear systems, in *Proceedings of the American Control Conference*, Albuquerque, IEEE, Piscataway, NJ, 1997, pp. 1804–1808.

Y. H. KIM AND F. L. LEWIS, *High-Level Feedback Control with Neural Networks*, World Scientific, River Edge, NJ, 1998.

P. V. KOKOTOVIĆ, Applications of singular perturbation techniques to control problems, *SIAM Rev.*, 26 (1984), pp. 501–550.

B. KOSKO, *Neural Networks and Fuzzy Systems*, Prentice–Hall, Englewood Cliffs, NJ, 1992.

B. KOSKO, *Fuzzy Engineering*, Prentice–Hall, Englewood Cliffs, NJ, 1997.

E. B. KOSMATOPOULOS, M. M. POLYCARPOU, M. A. CHRISTODOULOU, AND P. A. IOANNOU, High-order neural network structures for identification of dynamical systems, *IEEE Trans. Neural Networks*, 6 (1995), pp. 422–431.

Z. KOVACIC, V. PETIK, AND S. BOGDAN, Neural network based friction and nonlinear load compensator, in *Proceedings of the IEEE International Symposium on Industrial Electronics*, IEEE Computer Society Press, Los Alamitos, CA, Bled, Slovenia, 1999, pp. 157–161.

C. KRAVARIS AND R. A. WRIGHT, Deadtime compensation for nonlinear processes, *AIChE J.*, 35 (1989), pp. 1535–1542.

S. H. LANE AND R. F. STENGEL, Flight control design using non-linear inverse dynamics, *Automatica*, 24 (1988), pp. 471–483.

D. L. LAUGHLIN, D. E. RIVERA, AND M. MORARI, Smith predictor design for robust performance, *Internat. J. Control*, 46 (1987), pp. 477–504.

P. L. LEE AND G. R. SULLIVAN, A new multivariable deadtime control algorithm, *Chem. Engrg. Comm.*, 91 (1990), pp. 49–63.

S. LEE AND H. S. LEE, Design of optimal time-delayed teleoperator control systems, in *Proceedings of the IEEE International Conference on Robotics and Automation*, IEEE Computer Society Press, Los Alamitos, CA, 1994, pp. 3252–3258.

S.-W. LEE AND J.-H. KIM, Control of systems with deadzones using neural-network based learning control, in *Proceedings of the IEEE International Conference on Neural Networks*, IEEE Computer Society Press, Los Alamitos, CA, 1994, pp. 2535–2538.

S.-W. LEE AND J.-H. KIM, Robust adaptive stick-slip friction compensation, *IEEE Trans. Indust. Electron.*, 42 (1995), pp. 474–479.

J. LEITNER, A. CALISE, AND J. V. R. PRASAD, Analysis of adaptive neural networks for helicopter flight control, *J. Guidance Control Dynam.*, 20 (1997), pp. 972–979.

F. LEWIS, Nonlinear structures for feedback control, *Asian J. Control*, 1 (1999), pp. 205–228.

F. L. LEWIS, C. T. ABDALLAH, AND D. M. DAWSON, *Control of Robot Manipulators*, Macmillan, New York, 1993.

F. L. LEWIS, S. JAGANNATHAN, AND A. YESILDIREK, *Neural Network Control of Robot Manipulators and Nonlinear Systems*, Taylor and Francis, London, 1999.

F. L. LEWIS, K. LIU, R. R. SELMIC, AND L.-X. WANG, Adaptive fuzzy logic compensation of actuator deadzones, *J. Robotic Systems*, 14 (1997), pp. 501–511.

F. L. LEWIS, K. LIU, AND A. YESILDIREK, Neural net robot controller with guaranteed tracking performance, *IEEE Trans. Neural Networks*, 6 (1995), pp. 703–715.

F. L. LEWIS AND B. L. STEVENS, *Aircraft Control and Simulation*, 2nd ed., Wiley, New York, 2001.

F. L. LEWIS, A. YESILDIREK, AND K. LIU, Multilayer neural net robot controller: Structure and stability proofs, *IEEE Trans. Neural Networks*, 7 (1996), pp. 388–399.

W. LI AND X. CHENG, Adaptive high-precision control of positioning tables: Theory and experiment, *IEEE Trans. Control Systems Tech.*, 2 (1994), pp. 265–270.

J.-S. LIN AND I. KANELLAKOPOULOS, Nonlinear design of active suspensions, *IEEE Control Systems*, 17 (1997), pp. 45–59.

D. G. LUENBERGER, *Introduction to Dynamic Systems*, Wiley, New York, 1979.

M. MALEK-ZAVAREI AND M. JAMSHIDI, *Time-Delay Systems: Analysis, Optimization and Applications*, North-Holland, Amsterdam, 1987.

M. B. MCFARLAND AND A. J. CALISE, Adaptive nonlinear control of agile antiair missiles using neural networks, *IEEE Trans. Control Systems Tech.*, 8 (2000), pp. 749–756.

D. H. MEE, An extension of predictor control for systems with control time delays, *Internat. J. Control*, 18 (1971), pp. 1151–1168 (see also comments by J. E. Marshall, B. Ireland, and B. Garland in *Internat. J. Control*, 26 (1977), pp. 981–982).

W. T. MILLER III, R. S. SUTTON, AND P. J. WERBOS, *Neural Networks for Control*, MIT Press, Cambridge, MA, 1991.

K. S. NARENDRA AND A. M. ANNASWAMY, A new adaptive law for robust adaptation without persistent excitation, *IEEE Trans. Automat. Control*, 32 (1987), pp. 134–145.

K. S. NARENDRA AND K. PARTHASARATHY, Identification and control of dynamical systems using neural networks, *IEEE Trans. Neural Networks*, 1 (1990), pp. 4–27.

R. ORTEGA AND M. W. SPONG, Adaptive motion control of rigid robots: A tutorial, *Automatica*, 25 (1989), pp. 877–888.

Z. PALMOR, Stability properties of Smith dead-time compensator controllers, *Internat. J. Control*, 32 (1980), pp. 937–949.

J. PARK AND I.-W. SANDBERG, Universal approximation using radial-basis-function networks, *Neural Comput.*, 3 (1991), pp. 246–257.

J. PARK AND I. W. SANDBERG, Criteria for the approximation of nonlinear systems, *IEEE Trans. Circuits Systems*, 39 (1992), pp. 673–676.

K. M. PASSINO AND S. YURKOVICH, *Fuzzy Control*, Addison–Wesley, Menlo Park, CA, 1998.

M. M. POLYCARPOU, Stable adaptive neural control scheme for nonlinear systems, *IEEE Trans. Automat. Control*, 41 (1996), pp. 447–451.

P. RECKER, V. KOKOTOVIĆ, D. RHODE, AND J. WINKELMAN, Adaptive nonlinear control of systems containing a dead-zone, in *Proceedings of the IEEE Conference on Decision and Control*, IEEE Computer Society Press, Los Alamitos, CA, 1991, pp. 2111–2115.

G. A. ROVITHAKIS AND M. A. CHRISTODOULOU, Adaptive control of unknown plants using dynamical neural networks, *IEEE Trans. Systems Man Cybernet.*, 24 (1994), pp. 400–412.

N. SADEGH, A perceptron network for functional identification and control of nonlinear systems, *IEEE Trans. Neural Networks*, 4 (1993), pp. 982–988.

R. M. SANNER AND J.-J. E. SLOTINE, Stable adaptive control and recursive identification using radial Gaussian networks, in *Proceedings of the IEEE Conference on Decision and Control*, IEEE Computer Society Press, Los Alamitos, CA, 1991, pp. 2116–2123.

S. SCULLY, *Real Time Control with VRTX*, Internal report, Automation and Robotics Research Institute, University of Texas at Arlington, Arlington, TX, 1998.

D. R. SEIDL, S.-L. LAM, J. A. PUTMAN, AND R. D. LORENZ, Neural network compensation of gear backlash hysteresis in position-controlled mechanisms, *IEEE Trans. Indust. Appl.*, 31 (1995), pp. 1475–1483.

R. R. SELMIC, *Neurocontrol of Industrial Motion Systems with Actuator Nonlinearities*, Ph.D. thesis, University of Texas at Arlington, Arlington, TX, 2000.

R. R. SELMIC AND F. L. LEWIS, Deadzone compensation in motion control systems using neural networks, *IEEE Trans. Automat. Control*, 45 (2000), pp. 602–613.

T. B. SHERIDAN, Space teleoperation through time delay: Review and prognosis, *IEEE Trans. Robotics Automat.*, 9 (1993), pp. 592–606.

T. K. SHING, L. W. TSAI, AND P. S. KRISHNAPRASAD, *An Improved Model for the Dynamics of Spur Gear Systems with Backlash Consideration*, Technical research report, Institute for Systems Research, University of Maryland, College Park, MD, 1993.

J.-J. E. SLOTINE AND W. LI, Adaptive manipulator control: A case study, *IEEE Trans. Automat. Control*, 33 (1988), pp. 995–1003.

O. J. M. SMITH, Closer control of loops with dead time, *Chem. Engrg. Progr.*, 53 (1957), pp. 217–219.

Y. D. SONG, T. L. MITCHELL, AND H. Y. LAI, Control of a class of nonlinear uncertain systems via compensated inverse dynamics approach, *IEEE Trans. Automat. Control*, 39 (1994), pp. 1866–1871.

B. L. STEVENS AND F. L. LEWIS, *Aircraft Control and Simulation*, 2nd ed., Wiley, New York, 2001.

G. TAO AND P. V. KOKOTOVIĆ, Adaptive control of plants with unknown dead-zones, *IEEE Trans. Automat. Control*, 39 (1994), pp. 59–68.

G. TAO, AND P. V. KOKOTOVIĆ, Discrete-time adaptive control of systems with unknown deadzones, *Internat. J. Control*, 61 (1995), pp. 1–17.

G. TAO AND P. V. KOKOTOVIĆ, *Adaptive Control of Systems with Actuator and Sensor Non-linearities*, Wiley, New York, 1996.

M. TIAN AND G. TAO, Adaptive control of a class of nonlinear systems with unknown dead-zones, in *Proceedings of the IFAC World Congress*, San Francisco, 1996, pp. 209–214.

A. TZES, P.-Y. PENG, AND C.-C. HOUNG, Neural network control for DC motor micromaneuvering, *IEEE Trans. Indust. Electron.*, 42 (1995), pp. 516–523.

V. I. UTKIN, *Sliding Modes and Their Application in Variable Structure Systems*, Mir, Moscow, 1978, pp. 55–63.

L. VON BERTALANFFY, *General System Theory*, Braziller, New York, 1968.

L.-X. WANG, *Adaptive Fuzzy Systems and Control: Design and Stability Analysis*, Prentice–Hall, Englewood Cliffs, NJ, 1994.

L.-X. WANG, *A Course in Fuzzy Systems and Control*, Prentice–Hall, Englewood Cliffs, NJ, 1997.

L.-X. WANG AND J. M. MENDEL, Fuzzy basis functions, universal approximation, and orthogonal least-squares learning, *IEEE Trans. Neural Networks*, 3 (1992), pp. 807–814.

P. J. WERBOS, *Beyond Regression: New Tools for Prediction and Analysis in the Behavioral Sciences*, Ph.D. thesis, Harvard University, Cambridge, MA, 1974.

P. J. WERBOS, Neurocontrol and supervised learning: An overview and evaluation, in *Handbook of Intelligent Control*, D. A. White and D. A. Sofge, eds., Van Nostrand Reinhold, New York, 1992, pp. 65–89.

D. A. WHITE AND D. A. SOFGE, EDS., *Handbook of Intelligent Control*, Van Nostrand Reinhold, New York, 1992.

A. N. WHITEHEAD, *Science and the Modern World*, Lowell Lectures (1925), Macmillan, New York, 1953.

L. YAN AND C. J. LI, Robot learning control based on recurrent neural network inverse model, *J. Robotic Systems*, 14 (1997), pp. 199–211.

B. YANG AND H. ASADA, A new approach of adaptive reinforcement learning control, in *Proceedings of the International Joint Conference on Neural Networks*, Vol. 1, 1993, pp. 627–630.

P.-C. YEH AND P. KOKOTOVIĆ, Adaptive control of a class of nonlinear discrete-time systems, *Internat. J. Control*, 62 (1995), pp. 303–324.

H. YING, *Fuzzy Control and Modeling*, IEEE Press, New York, 2000.

Y. YOSHIMURA AND N. HAYASHI, Active control for the suspension of large-sized buses using fuzzy logic, *Internat. J. Systems Sci.*, 27 (1996), pp. 1243–1250.

Y. ZHANG AND Y. C. SOH, Robust adaptive control of uncertain discrete-time systems, *Automatica*, 35 (1998), pp. 321–329.

Index